The Observer's Pocket Series

FARM ANIMALS

The Observer Books

Natural History
- BIRDS
- BIRDS' EGGS
- BUTTERFLIES
- LARGER MOTHS
- COMMON INSECTS
- WILD ANIMALS
- ZOO ANIMALS
- WILD FLOWERS
- GARDEN FLOWERS
- FLOWERING TREES AND SHRUBS
- HOUSE PLANTS
- CACTI
- TREES
- GRASSES
- COMMON FUNGI
- LICHENS
- POND LIFE
- FRESHWATER FISHES
- SEA FISHES
- SEA AND SEASHORE
- GEOLOGY
- ASTRONOMY
- WEATHER
- CATS
- DOGS
- HORSES AND PONIES
- TROPICAL FISHES
- FARM ANIMALS

Transport
- AIRCRAFT
- AUTOMOBILES
- COMMERCIAL VEHICLES
- SHIPS
- MANNED SPACEFLIGHT
- UNMANNED SPACEFLIGHT
- BRITISH STEAM LOCOMOTIVES
- MOTORCYCLES

The Arts, etc.
- ANCIENT & ROMAN BRITAIN
- ARCHITECTURE
- CATHEDRALS
- CHURCHES
- HERALDRY
- FLAGS
- PAINTING
- MODERN ART
- SCULPTURE
- FURNITURE
- MUSIC
- POSTAGE STAMPS
- GLASS
- POTTERY & PORCELAIN
- BRITISH AWARDS & MEDALS
- EUROPEAN COSTUME
- SEWING
- TOURIST ATLAS G.B.

Sport
- ASSOCIATION FOOTBALL
- COARSE FISHING
- CRICKET
- GOLF
- MOTOR SPORT
- SMALL CRAFT
- SHOW JUMPING & EVENTING

Cities
- LONDON

The Observer's Book of
FARM ANIMALS

LAWRENCE ALDERSON,
M.A.(Agric.)

WITH 8 COLOUR PLATES
AND 109 BLACK
AND WHITE PHOTOGRAPHS

FREDERICK WARNE & CO LTD
FREDERICK WARNE & CO INC
LONDON : NEW YORK

© Frederick Warne & Co Ltd
London, England
1976

Library of Congress
Catalog Card No 76-2918

ISBN 0 7232 1555 3

Printed in Great Britain by
William Clowes and Sons, Limited
London, Beccles and Colchester
693.476

PREFACE

An increasing number of people are looking more to the countryside for their recreation, but it is not always easy for them to understand the many facets of rural life. A great deal has been done in recent years to help visitors to enjoy the pleasures of the country and to be more aware of the problems involved in farming and other rural occupations. This addition to the Observer series has been written to encourage and assist this process.

The domestication of animals and their use as farm livestock to produce food, clothing and other necessities has been an integral and important part of the development of civilizations and the progress of mankind. Over a period of several thousand years, different types of stock have been evolved initially by environmental influences; later the application of scientific selection methods by man resulted in the production of a large number of specialized breeds. The important international breeds are described and illustrated in this book, together with many other localized breeds which have played a more important role in the past.

Breeds of cattle, goats, horses, pigs and sheep are described in alphabetical order within each species, together with a summary of the development of poultry breeds. Both visual and performance characteristics are included, ensuring not only the easy identification of the various breeds, but also a deeper appreciation of their qualities and uses. A list of reference works is given on page 185 for those who wish to study farm livestock in greater detail.

ACKNOWLEDGEMENTS

The author and publishers wish to thank the following for their kind permission to reproduce photographs:

Livestock Improvement Service for Plate 1 (both), Plate 2 (middle and lower), Plate 3 (both), Plate 4, Plate 5, Plate 6 (upper), Plate 7 (lower); Jan Badenhorst for Plate 2 (upper), Plate 8, page 61 (lower), 127; Cleveland Bay Horse Society for Plate 6 (lower); E. J. Prince for Plate 7 (upper); Farmers Weekly for pages 25, 29, 33, 41, 45, 47, 49, 53, 56, 58, 61 (upper), 67 (lower), 69, 75 (upper), 77, 79, 80, 83, 85, 87 (both), 88, 93, 95, 97, 103, 105, 107 (lower), 109, 110, 113 (both), 115, 116, 121 (upper), 125 (upper), 131 (lower), 133, 137, 156, 157, 160-4, 166, 174 (lower), 176; British Farmer & Stockbreeder for pages 27, 31, 36, 39 (both), 43, 51 (both), 55, 65, 67 (upper), 75 (lower), 91, 98, 118, 121 (lower), 128, 131 (upper), 134, 138; Colbred Sheep Society for page 35; Farming News for page 71; Countrywide Livestock Ltd for page 72; Somerset Cattle Breeders for page 101 (upper); Avoncroft Cattle Breeders for page 123; Santa Gertrudis Breeders International for page 125 (lower); C. Hosegood for pages 139, 174 (upper); *The Observer's Book of Horses and Ponies* by R. S. Summerhays for page 141; Sally Anne Thompson for pages 143, 145, 146, 149, 154; Leslie Lane for pages 147, 153; American Quarter Horse Association for page 151; Chester White Swine Record Association for page 158; Poland China Record Association for page 165; Mrs D. A. Matthews for page 169 (upper); L. Avon for page 169 (lower); Mrs H. G. Awdry for page 171; A. Rice for 177, 178 (both), 181 (both), 182 (both), 183 (both), 184 (both).

CONTENTS

Preface	5
Acknowledgements	6
List of Colour Plates	8
Glossary	9
Metric Conversion Table	14
Introduction	15
Sheep	25
Cattle	93
Horses	141
Pigs	156
Goats	168
Poultry	173
Bibliography	185
Index	187

LIST OF COLOUR PLATES

Plate 1 Cotswold sheep in Gloucestershire parkland
Wiltshire Horn sheep shedding fleece in June

Plate 2 Soay ewes and lambs
Wensleydale ram in full fleece
White Shetland sheep

Plate 3 A typical English Longhorn bull
A herd of Shetland cows grazing moorland in Scotland

Plate 4 White Park bull and cow both 16 years of age

Plate 5 An imported Chianina bull at the Great Yorkshire Show

Plate 6 Belted Galloway cattle
A champion Cleveland Bay Horse

Plate 7 Team of Shire Horses ploughing
Middle White pig

Plate 8 'Iron Age' sow (Tamworth cross Wild Boar) with a litter sired by a Tamworth boar

Study this glossary before you read the text.

GLOSSARY

Anaplasmosis   A disease of cattle, carried by biting ticks.

Blood   Genetic influence. Bloodlines refer to groups of animals that belong to the same family, i.e. similar genetic makeup.

Bradford count   Refers to the fineness of wool, and indicates the theoretical number of hanks that can be spun from one pound (0·45 kg) of wool.

Britch   The thigh and twist region of the sheep. It produces the lowest quality wool.

Brisket   Chest.

Browse   To feed on the foliage of trees or shrubs.

Buds   Rudimentary horns. (See also **Scurs** and **Snags**.)

Coat (undercoat)   Hair. Some animals, especially those in cold climates have coats with a double layer for extra protection, and the inner layer is known as the undercoat.

Conformation   Shape.

Cotted   A condition of wool where it has become matted or felted while growing on the sheep. It is caused by insufficient wool oils being produced by the sheep, often due to sickness or injury.

Coupled   Is a term used to describe the length of the body of a horse. Close-coupled means short-bodied.

Crest   The muscular development of the neck as exhibited by bulls and stallions.

Crossbreeding (crossbred) The mating of animals both of which are purebreds but members of different breeds.

Dewlap Loose skin on the chest, particularly of cattle.

Dock To remove the tail at birth. This is common practice with sheep where a long tail causes disease and encourages parasites.

Draft After hill ewes have produced three crops of lambs they are transferred to the lowlands, and at this stage are known as draft ewes.

Dual-purpose Having two functions. This term is applied mainly to cattle breeds which are used both for meat and milk production.

Eye muscle The long muscle which extends from the base of the spine to the neck. Latin—*Longissimus dorsi*. It forms the major part of the meat in lamb or pork chops.

Feather The long hair which grows on the legs of some heavy horses below the knees and the hocks.

Feral Stock of domestic origin which has returned to the wild state, and descendants of such stock.

Finished A term used to describe the degree of fatness of animals intended for meat production. Finished indicates that the carcase will be of good quality.

Fly A parasite of sheep causing a condition known as fly strike where the grubs eat the wool and the flesh of the host.

Fold In the past some sheep were kept on a folding system, whereby they were restrained within enclosures or folds of hurdles which were moved to fresh ground at regular intervals.

Gaskin The second thigh. The part of the horse's leg immediately above the hock.

Gene A unit whose transmission determines the inheritance of a particular characteristic.

Hank 512 metres of wool.

Hefted A term applied to hill sheep which will remain in the area where they were born without being fenced in.

Heifer A young cow, usually less than two years old.

Heterozygous Genetically impure.

Hock Heel. The middle joint of the hind legs of animals.

Homozygous Genetically pure.

Hybrid A crossbred animal. Hybrid animals frequently exhibit hybrid vigour which results in greater efficiency and higher levels of production.

Inbreeding (inbred) The mating of closely related animals.

Kemps Thick, inelastic, hairy fibres in the fleece, which are interspersed with the wool fibres, and have a white, chalky appearance. They do not absorb dyes and are usually undesirable.

Lactation The yielding of milk. The period during which a female produces milk following birth.

Lamb (*verb*) To give birth (in sheep).

Lambing % The number of lambs produced by 100 ewes.

Landrace An indigenous breed adapted to its environment over a long period of time.

Ley Grassland which is ploughed up after a relatively short number of years.

Line-back An ancient colour pattern of cattle where a white stripe runs along the back and usually down the tail.

Marbled A quality of meat where a small amount of fat is interspersed with the lean meat, so improving its flavour and texture when cooked.

Mutation A mutation is a variation caused by a change in a gene or genes which is then passed on as an inherited characteristic.

Nick When two animals are mated and produce consistently good offspring, they are said to nick well.

Parturition Birth.

Phenotype Outward appearance. Visual characteristics.

Pigment Substance giving colour to tissues.

Polled Without horns.

Progeny Offspring.

Progenitor Ancestor.

Prolific Producing many offspring. Normally used with reference to sheep and pigs.

Purebred A purebred animal is one which is descended from animals which all belonged to the same breed.

Purled A type of wool where the locks are curled tightly into ringlets.

Roots Plants such as turnips used for feeding sheep.

Scurs Rudimentary horns. (See **Buds** and **Snags**.)

Shearling A sheep that has been shorn once, and is usually 15 to 27 months old.

Snags Rudimentary horns. (See **Buds** and **Scurs**.)

Soft Delicate. Not hardy.

Sprung Well curved (relating to the ribs of animals).

Staple A cluster or group of wool fibres naturally clinging together in the fleece. Staple length indicates the average length of the wool fibres.

Steer A castrated male (cattle).
Stratification A method whereby different breeds of sheep are crossed to combine their qualities to best effect.
Suckler cow A cow which rears her own calf, and is then used for beef production rather than milk production for human consumption.

Transmit Pass on.

Wedge-shape The ideal conformation of dairy cattle being deeper and broader in the hindquarters than in the forequarters.
Withers Shoulders.

METRIC CONVERSION TABLE

Metric and Imperial Equivalents Conversion Tables					
kg	lb	gal (approx.)	cm	in	hands (approx.)
1	2·2		5	2	
2	4·4		10	4	1
3	6·6		15	6	1·2
4	8·8		20	8	2
5	11·0	1	25	10	2·2
6	13·2		30	12	3
7	15·4		35	14	3·2
8	17·6		40	16	4
9	19·8		45	18·1	4·2
10	22·0	2	50	20·1	5
20	44·1	4	55	22·1	5·2
30	66·1	6½	60	24·1	6
40	88·2	8½	65	26·1	6·2
50	110·2	10½	70	28·1	7
60	132·3	12½	75	30·1	7·2
70	154·3	15	80	32·1	8
80	176·4	17	85	34·1	8·2
90	198·4	19	90	36·1	9
100	220·5	21	95	38·1	9·2
500	1 102·3	106	100	40·1	10
1 000	2 204·6	212	250	100·3	25
5 000	11 023·0	1 060	500	200·7	
			750	301·0	
			1 000	401·3	

INTRODUCTION

Since the ancestors of our present farm animals were first domesticated, about 7000 years ago, the evolution of the various breeds as we now know them has followed a continuous cycle of hybridization followed by the inbreeding of isolated pockets of animals to form a breed. In the past this pattern was relatively stable. Tribal migrations or military conquests were the main reasons for the movement and subsequent mixing of different types of livestock, but the periods of time between these upheavals were sufficiently long to permit well-defined breeds to be developed in each area. More recently, and especially in the 20th century, communications and transport facilities have enabled livestock of widely differing types from even remote areas to mix on a large scale, so that we are now in the midst of a phase of extensive hybridization that is likely to persist for some time.

It is possible to trace the evolution of some breeds through such a process over a period of many centuries. Before 1000 BC polled, that is hornless, cattle were found in the civilizations of Egypt and Mesopotamia. They were developed particularly in the area of advanced livestock breeding in Scythia. During the first millennium BC a general movement of tribal migrations from Scythia took polled cattle across Europe to Scandinavia. Here they became popular because they were more easily housed during the winter, and simpler to handle in a confined space. Some of these cattle had distinctive colour patterns, being white in colour with black points, that is nose, ears and feet. Even at that early time in history livestock breeders were attracted by animals of striking or unusual coloration, and these cattle became established in Sweden as a breed known as the Fjall. Several centuries later the Vikings carried their military conquests to many parts of Europe and took with them cattle to their newly acquired territories. In Great Britain polled cattle tend to be found in areas that were dominated by the Vikings, mainly in the east of the country. The British White in East Anglia is not only polled, but also has the same colour pattern as the Fjall.

Thus the British White breed can be traced back for more than 3000 years to an area which enjoys a much hotter climate than its present home. Evidence of this can be found in its white coat and pigmented skin, which are normal features in the native cattle of the eastern Mediterranean region and of the Indian sub-continent. Finally, the British White is now being used as one of a trio of breeds to create a new master breed for the North American range country, the region from the mid-West to the foothills of the Rocky Mountains.

The wild cattle, from which modern domesticated cattle are descended, are now extinct. It is generally accepted that the auroch is the progenitor of modern cattle, and the last specimen died in the forests of Poland in 1627. An attempt has been made by Professor Heinz Heck, in the Munich Zoological Gardens, to recreate the auroch by mixing selected breeds of cattle. He used Steppe cattle, Highlands, Alpine cattle and Friesians. The resulting animals resembled the original auroch very closely, and the progeny they produced were true to type. Not only were they coloured correctly, showing the line-back pattern, with the bulls darker than the cows, but also their temperament very quickly became typical of wild animals.

The wild progenitors of most other species can still be found. Goats are unusual because there has been considerable interbreeding with wild and domesticated stock, and the feral herds in Great Britain probably have evolved in this way. However, on the mainland of Europe wild boar still roam in many forests. An adult boar weighs 170 kg and the length of the head and body is 140 cm. They have been crossed with domestic pigs, and the hybrid piglets carry on their bodies horizontal stripes of black and red like those of wild boar. Four types of wild sheep still exist, and three of these probably contributed to the formation of domestic breeds. The exception is the Big Horn of North America. Influences of both the Moufflon of Corsica and the Urial of Asia can be traced in the Soay, while the heavily-spiralled horns of the Argali recur in the blackfaced mountain breeds of the Pennines of England.

Although many breeds of livestock have become extinct during the long process of domestication and improvement, it is still possible to trace the stages of progressive development through different breeds. This is particularly true of sheep where the absence of restricting legislation has permitted the survival of a wide variety of breeds. Improvement of the quality of livestock is achieved either by natural long term evolution, which enables animals to adapt themselves to their environment, or by artificial selection by man to produce higher yielding types, or by a combination of both methods.

The primitive breeds of sheep are largely those which have evolved with minimum human interference. The most primitive is the Soay, which resembles closely the wild Moufflon, and which is a living link between wild and domesticated sheep. The next stage of development is represented by the seaweed-eating North Ronaldsay sheep of the Orkney Islands, which are still small and fine-boned, and have a short tail and multi-coloured wool. After this stage some sheep were bred for wool production, while in other cases meat production or yield of milk was more important. Shetland sheep, which are closely related to the North Ronaldsay, can still be found in a variety of colours, but white is more common and the wool is noted for its softness and fineness. They are hardy sheep and can survive in the harsh climate of their native islands.

Breeds of sheep which are kept for the production of meat or milk owe more of their present characteristics to the ideals and objectives of sheep breeders. The tan-faced sheep of Britain, which were developed from the Soay, were represented in the south-west of England by the Portland. The meat of Portland sheep in the past has enjoyed a good reputation and is still noted for its quality and flavour. The final stage of improvement is the Dorset Horn, which is derived from the Portland, and which has compared very favourably in trials in Ireland to demonstrate the carcase quality of various breeds. Although sheep were used first for milk production, and in many parts of the world this is still their primary function, in northern Europe it applies only to specialist breeds, most of which are based on the East Friesland. The Milksheep of Britain have been specially selected and highly developed to achieve extreme yields of milk production. Other breeds which have high levels of production are the result of artificial selection programmes, and tend to be more reliant than primitive breeds on artificial aids such as housing and extra feeding.

Where a breed of livestock has been exploited for one particular characteristic it requires a higher standard of management, and other qualities are likely to have deteriorated. Thus Finnish Landrace sheep are prolific, but they produce a carcase of poor quality and, as housing is normal practice in Finland, they cannot tolerate wet climates. In the same way, Landrace pigs have been bred for bacon carcase quality and for long bodies to such a degree that their legs are frequently unable to support their elongated frame, while Pietrain pigs are very heavily muscled, but their meat is often of poor quality being pale and watery. Among dairy cattle the Jersey produces the richest milk, but it is of little use for beef production, while one of the world's leading beef breeds, the Charolais, has such large hindquarters that the cows frequently experience difficulty in giving birth to calves, which already show this characteristic when born.

Breeds of livestock can be classified in several different ways. Most systems use commercial qualities, but sheep may be categorized on the basis of some distinctive visual characteristic such as the wool or tail. Wild sheep have a short, thin tail, but modern, highly developed breeds would have long woolly tails if they were not docked at birth. The docking operation is necessary with long-tailed sheep that are kept under normal farming conditions. Tails may be short or long, bald or woolly, thin or fat. Fat-tailed sheep are found in hot climates, and the store of fat in the base of the tail serves the same purpose as the hump of the camel. Wool types are represented at one extreme by the fine wool of the Merino and at the other by the hairy sheep of Africa and Asia; intermediate types include coarse wool, long wool, short wool and kempy wool. Coarse wool and kempy wool provide better protection against adverse climatic conditions and are found on hill breeds. Most of the heavily-fleeced longwool breeds which originated in the British Isles are derived from stock introduced by the Romans.

Sheep farming in Britain is based on the system of stratification in which breeds with differing characteristics are crossed to combine their qualities. Hardy hill breeds are the starting point of this system. Breeds such as the Swaledale, Cheviot, Scottish Blackface, Welsh Mountain and Exmoor Horn can tolerate their harsh environment during the early years of their life, but later they are transferred to lower land, where they are mated with the rams of prolific breeds which transmit either high growth rate or high milk yield to their hybrid progeny. The most important breeds in this category are the Bluefaced Leicester, Border Leicester, Colbred and Milksheep, and their progeny out of hill ewes are known as the Mule, Halfbred, T. C. hybrid and Milksheep hybrid respectively. These hybrids combine the qualities of both parent breeds, and are mated to a Down ram to breed lambs which produce a high quality carcase. The Down breeds include the Suffolk, Dorset, Hampshire, Shropshire, Oxford, Ryeland and Southdown, and have been highly developed for their meat-producing ability. Some breeds do not fit into the stratification system and are kept more in purebred flocks. They are known as grassland breeds, and include the Clun Forest, Kerry Hill, Devon Closewool and Dorset Horn.

Among British breeds of sheep there are four main categories, namely hill, crossing, Down and grassland. In other countries the classification is not so well defined, and planned cross-breeding is less evident. Both in North America and on the mainland of Europe there is greater emphasis on purebred stock, but Australia has evolved her own system of stratification with the Merino, Leicester and Dorset Horn.

Cattle are kept mainly for three purposes: work, meat and milk. Some breeds fulfil all three roles and are thus triple-purpose, but the number of cattle used as draught oxen is decreasing rapidly so that most cattle breeds can be classified as either beef, dairy or dual-purpose. Beef breeds include the Highland, Hereford and Aberdeen Angus from Great Britain, the Charolais from France, the Chianina from Italy and the Santa Gertrudis from North America, but there is a wide variation in type from the small, squat, compactness of the Aberdeen Angus to the long-legged, lean, massiveness of the Chianina, and from the shaggy hardiness of the Highland to the rounded muscularity of the Charolais grazing the lush pastures of the Loire basin.

Numerically, dairy cattle are dominated by the Friesian and Holstein, and as long as the financial returns to dairy farmers favour quantity rather than quality, it is unlikely that breeds such as the Jersey or Guernsey, which produce rich milk, will regain any of the ground which they have lost to the Friesian. During much of the 20th century the Friesian has also successfully replaced dual-purpose breeds in many areas. In Britain the Shorthorn has become a minority breed within 50 years of being the dominant breed. However, the emphasis is now changing as the leading dual-purpose breeds, the Simmental and the Meuse-Rhine-Ijssel, are enjoying increased popularity, and even the Friesian is showing signs of becoming a dual-purpose breed.

The common ancestor of all breeds of horse was *Equus przevalskii*, which existed during the middle and late Tertiary period and persisted much later as the wild horse of the Mongolian Steppe. It gave rise to two main lines of development. The first was a direct descendant and has a black dorsal stripe and an upright mane, and stands 12·1 to 14·1 hands high. The Tarpan was the basis of the second type. It is now extinct, but most of the Oriental and primitive pony breeds were developed from it. It also had a black dorsal stripe and a withers height of 13 hands. The evolution of modern breeds from these primitive types was influenced to a large degree by the environment in which they lived. In hot climates the 'hot-blood' breeds were elegant creatures with a thin skin and a smooth coat. They were used for racing as early as 1000 BC, and were the main ancestor of the Thoroughbred. The Arab is the most typical modern representative of this group. In cold climates the animals had a smaller stature and a thicker coat, and most of the modern pony breeds, such as the Exmoor, are included in this category.

Meanwhile, a separate and distinct type had evolved in Europe during the Quaternary period. This was the heavy prehistoric horse which grazed the rich marshlands and valleys of central and northeastern Europe. In contrast to the light, fast horses found in hotter climates, these heavy, coarse creatures were known as 'cold-bloods'. They grew to a great size and had a passive temperament. The oldest representative of this group is the Ardennes of northern France and Belgium, which was the immediate ancestor of the modern Belgian. Most breeds are now a mixture of hot and cold blood types. Thus the Percheron is a heavy draught breed, but it has been refined by the introduction of Arab blood, while the Dales Pony has been influenced by crosses with the Clydesdale, so that it is sometimes classified as a miniature cart horse.

The domesticated pig is descended from two wild

ancestors, the wild pig of Europe, *Sus scrofa*, and the wild pig of eastern Asia, *S. vittatus*. Until the late 18th century the British pig was a dirty yellow colour with darker markings and a coarse coat, and was a direct descendant of the European wild boar. It was a massive animal which occasionally weighed as much as 550 kg. In 1770 Chinese pigs were imported into Britain. They were descended from the Asian wild boar, and had small, broad heads with prick ears, short, fat bodies and short legs. There were both black and white types. Previously Asian pigs had found their way to Italy and from there the black Neapolitan came to England. The only other importation was the Red Barbadan which gave its colour to the Tamworth. The influence of Chinese pigs was felt throughout Britain and is most evident in early maturing breeds such as the Berkshire and Middle White. The Landrace of northern Europe and related breeds owe nothing to Chinese influence, being directly descended from the European wild pig. They are long-bodied, lean and late maturing, and have lop ears. Between these two extremes most other breeds combine the qualities of the two parent types in various permutations.

The whole process of domestication and livestock improvement has been one of increasing control by livestock farmers resulting in an ever-widening gap between the requirements of domestic animals and their natural environment. The primitive Soay sheep thrives and reproduces on its island home in the Outer Hebrides without human assistance, but its descendant—the Finnish Landrace—requires housing in the winter, receives assistance at parturition, has feet which are susceptible to disease, and wool which does not shed water. The justification for this is that the Finnish Landrace produces twice as many lambs as the Soay, and high performance standards have received top priority in recent decades. However, under some conditions more efficient production can be achieved by using breeds that are adapted to the prevailing natural conditions, thus eliminating expensive and unnecessary artificial aids. The Soay sheep are an excellent example of this, and in the same way Tamworth pigs living in rough scrubland may be more efficient than Landrace pigs in controlled environment housing, while the Texas Longhorn is once again becoming popular because of its ability to utilize marginal land of low fertility.

In some parts of the world livestock farming has taken this argument to its logical conclusion. As wild animals are best adapted to the natural environment, they may be the most efficient producers of meat. In tropical areas of Africa, eland and oryx antelope can produce a valuable output without pharmaceutical aids to keep disease and parasites at bay. Similar results have been obtained in North America with bison, while in New Zealand and Scotland the red deer has been chosen for farming experiments. These wild animals of today may be the domesticated farm animals of the future.

SHEEP

Bleu du Maine

During the 19th century in France it was the fashion to import Durham bulls and Leicester rams, and many French breeds were influenced by these importations. In certain instances other imported breeds played a significant role, and the Bleu du Maine owes part of its ancestry to the Wensleydale. Although the Bleu du Maine was derived mainly from the use of Leicester rams on the native sheep, it has inherited from the Wensleydale deep blue pigmentation of the face and ears, size, strong chest and shoulders, and prolificacy. The breed was recognized officially by the Ministry of Agriculture in 1948. It is

Bleu du Maine ram

still fairly localized in France, in the area just west from Le Mans, but it has been exported to a few countries, of which Ethiopia is the most important.

The Bleu du Maine is a large sheep; mature ewes weigh about 85 kg and produce a fleece weighing slightly more than 4 kg. It has a dark blue face clear of wool. The ears are long with strong pigmentation and the eyes are prominent. Despite the selection of breeding animals with good blue colour, variation still occurs from black, at one extreme, in common with the Wensleydale, to red at the other. It is significant that another breed, the Rouge de l'Ouest, has sprung from the same native stock as the Bleu du Maine.

As regards production, the Bleu du Maine is one of the best French breeds. It achieves a high level of prolificacy with the best flocks averaging more than 200% lambing. Although the lambs tend to be rather delicate when young, they are capable of making rapid weight gains comparable to the Texel, and 35 grams per day faster than the Île de France. Bleu du Maine ewes milk reasonably well. They are classed as a grassland breed. They have steadily increased in numbers during the present century, rising from about 30,000 sheep in 1930 to a present total population of about 150,000 sheep. The average flock contains about 60 ewes.

Bluefaced Leicester

During the early years of the present century, breeders of the whitefaced Border Leicester sheep in the north of England realized that strains of this breed which had a darker face sired progeny of better quality. Consequently they selected and bred the new type which became known as the Bluefaced or Hex-

Bluefaced Leicester ewe

ham Leicester. It seems likely that either the Wensleydale or the Teeswater made some contribution to the new breed, which was found mainly in the Tyne valley and on the lower western slopes of the Pennines in Cumbria. This breed had emerged and become established as a distinct type by 1914, but a Breed Society was not formed until 1962.

The Bluefaced Leicester is a long sheep with a large body. It is not unusual for ewes to produce triplets with a total birth weight in excess of 16 kg. Sheep of this breed have a bold head, with a distinct Roman nose, and large semi-erect ears. The skin is pigmented, especially on the ears, over the top of the head and down the back. These sheep are fine-skinned and produce a relatively light fleece of tightly purled lustre wool, which does not provide adequate protection in a cold, wet winter. Delicacy is a fault of the pure sheep, although their crossbred offspring show considerable hybrid vigour and hardiness.

This is one of the few breeds that has been progeny tested on an extensive scale. Bluefaced Leicester rams are selected for pure breeding according to the quality of their crossbred daughters. The most important progeny is the Mule ewe, which has proved itself to be the outstanding commercial sheep readily available to sheep farmers in Great Britain. In large scale trials it achieved the highest levels of prolificacy and profitability.

An adult Bluefaced Leicester ewe weighs about 86 kg and averages over 200% lambing, with the more productive strains reaching 250% lambing. Lamb growth rates are high, frequently exceeding 450 grams per day, but the sheep require a high level of feeding in order to maintain this standard of production. There are at present about 400 registered flocks containing about 4000 breeding ewes, but the breed is currently extending its territory. It is spreading into the Border country at the expense of the Border Leicester on account of its superior prolificacy, and is replacing the Teeswater in the north of England because its lambs are more active at birth.

Border Leicester

In 1767 the Culley brothers introduced Dishley Leicester sheep into the Scottish Border country. Much controversy surrounds the subsequent breeding of these animals. It is claimed that the Border Leicester evolved from these original animals simply by selection, but most authorities agree that this breed owes a significant part of its ancestry to the Cheviot. By the

Border Leicester ram

end of the 18th century, the Border Leicester was being used to cross onto hill Cheviot ewes to produce the Scotch Halfbred ewe, which by 1850 was the principal sheep export from the Border hills. It was not until 1869 that the Highland Agricultural Society provided separate classes for the English and Border Leicester, and a Breed Society for the latter was formed in 1898.

The Border Leicester was developed as an early maturing breed suitable for crossing with thin-fleshed and slow-feeding hill breeds. It is crossed mainly with the Scottish Blackface, Cheviot and Welsh Mountain, to produce Greyface, Scotch Halfbred and Welsh Halfbred breeding ewes respectively, and it is still the most widely used crossing ram. It is used as the control breed in trials to assess the merits of new breeds, and results suggest that the Border Leicester may become outclassed. In France it was unable to compare with the Romanov, while in Great Britain it is being superseded by the Bluefaced Leicester.

The Border Leicester is long-bodied, with a well-developed chest and the brisket carried well forward. Its white, clean face with large erect ears, and the proud arch of its head and neck make it a very distinctive and attractive sheep. Mature ewes weigh 80–85 kg and produce a 4 kg fleece of lustre wool with a staple length of 20 cm and a Bradford Count of 40–46. The breed is now less prolific owing to undue emphasis on success in the showring, and now averages about 170% lambing.

There are about 400 registered flocks of Border Leicester sheep in Great Britain containing about 10,000 breeding ewes. The breed has been exported and is represented particularly strongly in Australia, where there are more than 1000 flocks.

Cheviot

Two hundred and fifty years ago Scotland was populated by a primitive, tan-faced type of sheep still found in isolated areas and on the islands. During the 18th century the northward sweep of the Cheviot and the Scottish Blackface completely changed this long established pattern. A struggle for territory between the Cheviot and Scottish Blackface breeds, known respectively as 'long sheep' and 'short sheep', took place. When the battle, which reached its peak in the first half of the 19th century, had subsided the Blackface occupied the Scottish Highlands, dividing the Cheviot strongholds of the Border country and Caithness. In the latter area a distinctly different type was developed, known as the North Country Cheviot. Meanwhile a third type known as the Sennybridge

Cheviot resulted from the importation of Cheviots into Glamorgan around 1820–30.

Originally the Cheviot was a small fine-boned sheep, but from 1761 introductions of Lincoln and Leicester blood improved the conformation and carcase, strengthened the light forequarters, and increased the weight. If careful attention is not paid to breeding, the Cheviot can revert to the tan-faced type and colouring, with red kemp in the fleece. The original sheep were horned, and rams with horns are not uncommon even today. At the end of the 18th century some Ryeland and Merino rams were introduced to improve the quality of the wool.

In the 1870s and 1880s undue emphasis was placed on show points and, although the breed was still prolific, it became large and soft, with a higher level of mortality. It was at this time that the Blackface proved that it produced better on the harsher 'black'

North Country Cheviot ewe

heather land, while the Cheviot required 'white' grass moor in order to show its superior breeding potential. The Breed Society was formed in 1891 and a separate Society for the North Country Cheviot was established in 1912.

The modern Cheviot is a medium-sized sheep, which is suitable for the production of light lamb with a carcase weight of 12·5–14·5 kg. It has a white face and legs, clear of wool, with ears held erect to give a distinctive alert appearance. On lowland farms mature ewes average about 150% lambing, and produce 2 kg of wool with a staple length of 10 cm and a Bradford Count of 50–56. The North Country Cheviot is larger, with ewes weighing 73·5 kg.

The total population of Cheviot sheep in Great Britain numbers rather less than 1,000,000. Draft ewes have been crossed with the Border Leicester since the end of the 18th century to produce the Scotch Halfbred, for many years the most popular crossbred ewe in the lowlands. The Cheviot is found in many other countries. It was first exported to North America in 1825. In Norway 20% of the sheep are Cheviots, while in New Zealand a new breed, the Perendale, has been developed in the last 25 years by crossing the Cheviot and Romney breeds.

Clun Forest

The Clun Forest was derived from a variable assortment of local types in the West Midlands, which belonged to the group of heath sheep that were related to the Hereford, ancestor of the modern Ryeland, and to the unimproved Radnor. Later it received an additional contribution from the Shropshire, and there was considerable variation within the breed at the beginning of the present century, when it was not in-by 1965 there were over 1100 registered flocks. This population explosion was due to the abandonment of the old folding system, and to the suitability of the Clun Forest to the ley system that took its place. At present the breed numbers about 40,000 breeding ewes, spread throughout the country. It has been exported to several European countries and has established a firm base in France. In Great Britain the rapid expansion of the breed again resulted in marked variation both in appearance and performance, but the best strains were sheep of high productivity and these were used in the creation of both the Cambridge and the Colbred.

Colbred

The sheep industry of the 20th century has been so strongly steeped in tradition that there has been aversion to change. The first man to disturb this inertia was Oscar Colburn of Crickley Barrow in Gloucestershire, who created a new breed called the Colbred. His objectives were to breed a sheep that would sire prolific and milky crossbred ewes out of native hill ewes. He produced the Colbred from four breeds, namely the East Friesland, the Dorset Horn, the Border Leicester and the Clun Forest. The inclusion of the East Friesland was particularly important for its prolificacy and very high yield of milk, and it was imported in 1957.

Oscar Colburn estimated that it takes 10 generations, or about 20 years, to establish a new breed. During that time the sheep are selected carefully for fertility and milking capacity. This breed has now achieved a reasonable degree of structural uniformity, and has been exported. All breeders are required to keep performance records.

Clun Forest ram

cluded in the majority of surveys of sheep breeds. The Clun Forest is a compromise between hill sheep (Welsh heath) and Down breeds (Shropshire), and has been established as a grassland breed.

It has a dark brown face with a covering of wool extending over the poll to the forehead. The ears are short and pricked, giving an alert appearance. Adult ewes in the uplands weigh about 55 kg and average 159% lambing under lowland conditions. The average fleece weight is 2·75 kg, with a staple length of 9 cm and a Bradford Count of 48–56. The Llanwenog is a very similar breed, mainly localized in south-west Wales. It is more prolific in its native area, and arose from a cross at the end of the 19th century between the Shropshire Down and the Llanllwni, a horned blackfaced breed.

The Breed Society for the Clun Forest was formed in 1925. By 1945 there were 100 registered flocks, and

The Colbred is a long sheep, with fairly long legs and hips carried well back to permit easy lambing. It has a white face, clear of wool, pink skin and nostrils, and large pinkish ears. Adult ewes weigh 70–75 kg and produce about 3·5 kg of silky wool. They rear about 1·8 lambs per ewe, and yield 200 kg of milk, which is appreciably more than most British breeds produce.

The Colbred can be used as a pure breed, mated to Down rams, to produce lambs suitable for the supermarket trade, but its main purpose is to cross with hill ewes for the production of crossbred breeding ewes. It transmits a higher yield of milk and earlier sexual maturity than other crossing rams, but it is slightly inferior to the Bluefaced Leicester for prolificacy and growth rate. Numbers of the breed are limited at this time, but it is being increased in conjunction with a controlled programme of progeny testing rams.

Colbred ram

Cotswold

In medieval times the Cotswold hills were famous already as a centre of the English wool trade. The name itself derived from the Saxon words 'cote' (sheepfold) and 'wold' (bare hill). Originally the native sheep of the area were of the ancient middle-wool type, but continual crosses with the unimproved Leicester changed its appearance until by the middle of the 17th century it had become a typical member of the longwool group. The massive longwool breeds have evolved generally in rich lowland areas, and it is unusual to find such a breed on the thin brashy limestone soil of the Cotswold hills. The Cotswold sheep remained a relatively old-fashioned type of longwool, being late maturing, long and broad backed, and having relatively large hindquarters. However, crosses with the Midland longwools moved

Cotswold ewe

the weight to the forequarters and the Cotswolds grew coarse and over fat, especially at higher weights.

The breed was at the height of its fame about 1850 when as many as 5000 rams were sold or let in a season. Since that time it has declined in the face of competition from the Down breeds. The Breed Society was formed in 1892, but by 1950 the breed was kept alive only by the persistence of one man, Mr W. Garne of Aldsworth in Gloucestershire. At present the breed is increasing slightly, but is still very limited in numbers.

The Cotswold has a white or grey face with a well-developed forelock. An adult ewe weighs about 85 kg and produces a fleece of about 5 kg. The wool is easily cotted and has a staple length of 30 cm with a Bradford Count of 38–44. Several flocks of Cotswold sheep are lambed in late January and February, and are reasonably prolific compared with other longwool breeds. A Cotswold ewe has been known to produce sextuplets.

The Cotswold has contributed to the formation of other breeds. In conjunction with the Hampshire Down it formed the Oxford Down, and crossed onto the native Marsh sheep of Germany, it produced the German Whiteheaded Mutton (Oldenberg).

Derbyshire Gritstone

The Derbyshire Gritstone has existed for many years as a distinct type of sheep in the Peak District of the Pennines. It owes the major part of its ancestry to the blackfaced, horned group of mountain breeds, but at various times in the past Leicester and Limestone breeds have been introduced and have contributed to the development of the Derbyshire Gritstone. A Breed Society was founded in 1906 and the first volume of the Flock Book was published in 1907.

The Derbyshire Gritstone is the biggest of the hill and mountain breeds. Unlike other related breeds it is polled in both sexes, and the face and legs are mottled black and white. It produces the highest quality wool of all the mountain breeds, with a Bradford Count of 50–56. An average fleece weighs about 2·25 kg with a staple length of 18 cm, and it is used for the manufacture of high quality hosiery wools.

Unlike the other Pennine breeds, the Derbyshire Gritstone is not widely used as the dam of crossbred breeding ewes, but it is more often crossed with Down rams to produce good quality lamb.

The breed is not strong numerically and remains localized mainly in its area of origin, although the transfer of Derbyshire Gritstone to Wales may have contributed to the formation of the Beulah Specklefaced.

Devon Longwool

The Devon Longwool is a member of the polled longwool group of breeds which is represented additionally in the south-west of England by the South Devon and Dartmoor. These three breeds all originated from the native short wool and middle wool types, which were whitefaced and horned. An early longwool cross started the transformation of these sheep which became known collectively as Bamptons. The process was completed by the introduction of Dishley Leicester rams. The Devon Longwool was influenced most by these crosses, while the Dartmoor is closer to the original type and the rams are horned occasionally. A Breed Society for the Devon Longwool was formed in 1898 and it is numerically stronger than the South Devon and Dartmoor, for which Breed

(*above*) Derbyshire Gritstone ram

(*below*) Devon Longwool ram

Societies were formed in 1904 and 1909 respectively.

The three breeds are of similar type, being massive sheep with a large head and heavy bones. The face is white with black spots on the ears, and the head well covered with wool. The body is covered with a coat of long, curly, lustre wool of wide staple and is free from kemp. The average fleece weight is 6·5 kg with a staple length of 20–30 cm and a Bradford Count of 32–36; mature ewes weigh about 85 kg. The South Devon is slightly heavier, while the Dartmoor is smaller, shorter in the leg and produces a heavier fleece of higher quality wool.

The influence of Dishley Leicester blood was less strong in the South-West, so that the longwool breeds of this area retained some of the prolificacy and milking ability of the native sheep, although not to the same degree as in the Teeswater. Despite this, the Devon Longwool achieves only 125–133% lambing, and various breeders are now embarking on an improvement programme. They are using rams of high performance breeds such as the East Friesland, which transmit high milk yield and prolificacy, and which will produce an 'improved' type of sheep in the South-West capable of achieving 200% lambing, with sufficient milk to fatten triplets, and an average fleece weight of about 5 kg.

Dorset Down

The development of the Dorset Down breed was contemporary with that of the Hampshire Down. It was based on local, mainly polled, sheep in Dorset, which were crossed first with Southdown and then with rams of the newly emerging Hampshire breed. Thus the Dorset Down was a compromise between the carcase quality and quick maturity of the Southdown,

Dorset Down ram

and the high growth rate of the early Hampshire.

The Breed Society was formed in 1906, later than the other Down breeds, but today it is numerically superior to all the Down breeds except the Suffolk, having about 80 registered flocks with 8000 breeding ewes. While it does not possess the out-of-season lambing ability of its close neighbour, the Dorset Horn, it does have a relatively long breeding season, and averages 135% lambing. It is used mainly on crossbred ewes for the production of quality lamb.

Mature ewes weigh on average 70 kg, so that it is one of the smaller Down breeds, and this is reflected in the lower growth rates of its progeny. On the other hand progeny of the Dorset Down have excellent conformation and carcase characteristics, with a large eye muscle, although the proportion of fat in the total carcase tends to be high.

The face is brown in colour and is well woolled down to the eyes. The wool is of good quality being used in the manufacture of high quality hosiery and fine knitting wools, and in the manufacture of paper felts used in the production of bank notes. An average fleece weighs 2·75 kg, with a staple length of 6·5 cm and a Bradford Count of 56–58. The quality of its wool has been partly responsible for the export demand for Dorset Down sheep, which are now found in many countries, particularly in Australasia.

Dorset Horn

Although many breeds in various parts of the world have the ability to lamb at any time of the year, the only breed in Great Britain which lambs out of season is the Dorset Horn and its sub-variety the Poll Dorset. This characteristic was recorded in the Dorset Horn as early as the 17th century, and is now a major factor contributing to the current success of the breed. The sheep inhabiting the south-west of England belonged originally to the tan-faced group, which is represented now by the Portland. The Dorset Horn emerged from this group as a result of the crossing of Merino-type sheep with the tan-faced sheep. There has also been a limited introduction of Southdown rams, but they exerted little influence on the development of the Dorset Horn.

The Dorset Horn was exported widely to North America, South Africa and Australasia. It was in Australia that the Poll Dorset was developed, by the use of Ryeland rams, and was exported to England and from there to several countries in the Mediterranean region. In Great Britain the breed is still found mainly in its native county, but several leading flocks have been established further afield. The Dorset

Dorset Horn ram

Horn is horned in both sexes. Frequently the horns of the rams grow too close to the face and need to be sawn. The face is white with pink nostrils and has a medium covering of wool over the poll. Mature ewes weigh 70 kg and average about 140% lambing, but some highly productive strains have been developed which lamb at over 200% and which usually weigh nearer 80–85 kg. Individual ewes have achieved a lifetime production of 30 lambs. The Poll Dorset is identical to the Dorset Horn in all points except that it is hornless.

The Dorset Horn produces about 2·75 kg of remarkably white wool, which is free from kemp, and which has a Bradford Count of 54–58. The staple length is 9 cm, and the wool is used for hosiery, dress fabrics and fine tweeds. Trials in Ireland have indicated that the Dorset Horn can be used also as a meat ram to sire crossbred progeny, with a high proportion of lean meat in the carcase and a high proportion of the weight carried in the expensive cuts of the hindquarters.

This breed can be used to great effect to cross with primitive breeds which are deficient in carcase quality. In particular it nicks well with the Finnish Landrace and the old-fashioned parkland type of Jacob. The Dorset Horn was used also in the breeding of the Colbred and, crossed with the Blackheaded Persian, produced the Dorper.

East Friesland

It is probable, when sheep were first domesticated, that they were developed as a milch animal, the lambs being weaned after a few weeks and the milk thereafter being used for human consumption. This practice is still common in many parts of the world, including Europe, where the production of milk is more important than the yield of either meat or wool. Milking ability has been developed to a high degree in several breeds of sheep. The milk of the Lacaune, which is found in the Massif Central of France, is used in the manufacture of Rocquefort cheese. The Awassi was the milk breed selected and improved by the Israelis, while the Chios is a high yielding breed found on the Greek island of that name and the adjacent mainland of Turkey.

The most highly developed breed of milksheep is the East Friesland. It is significant that the world's leading breed of dairy cattle had its origins in the same area of northern Holland and the adjoining province of Germany. The East Friesland was based on the old Marsh sheep of Holland, which was modified to a small degree by crosses with imported Leicester rams. However, it is unlikely that the Dishley

East Friesland ewe

Leicester rams, which lost both prolificacy and milk yield as a result of Bakewell's policies (see p. 57), played a significant role in the development of the East Friesland.

The modern East Friesland is a large, slim animal, with mature rams occasionally weighing up to 150 kg. It has a white face with pink nostrils and long ears. It is long-bodied and fine-boned, and the wide pelvis with the hips carried well back permit easy lambing. The wool is white and fine, and an average fleece weighs about 5 kg. Perhaps the most distinctive physical feature of the breed is its thin bald tail. It is late maturing and finishes at high weights, but the ewe lambs and rams are precocious and breed readily in their first year.

The East Friesland has been milk recorded for many years and has averaged a yield of 450 kg at 7%

butterfat in a 200-day lactation. By selective breeding the average yield has now been raised to 600 kg. This level of production has caused the breed to be known as the 'small-holders' cow'. It is also a prolific breed averaging over 200% lambing, and has been used to raise the productivity of most other breeds of milk-sheep. It was used by Oscar Colburn in the development of the Colbred. Its main function at present in Great Britain is to improve the native longwool breeds, especially those with low performance potential in the southern part of the country.

The East Friesland is now relatively low in numbers, but it is making a significant contribution to many current sheep improvement programmes.

Exmoor Horn

Before Ellman and Bakewell (see p. 57) began their sheep improvement programmes, the south-west of England was populated by the local variety of the Whitefaced Horned group of sheep, of which the modern representative is the Exmoor Horn. When the Breed Society was formed in 1906 the Exmoor Horn was a relatively unimproved sheep that had evolved in the inhospitable climate of its native moorland at heights up to 300 metres. It is not as hardy as many of the other hill breeds, but it possesses a dense fleece which effectively sheds the rain.

In recent decades the emphasis of improvement has been placed more on fleece quality and early maturity. The wool is soft with a staple length of 9 cm and a Bradford Count of 50–56. An average fleece weighs 2·5 kg. In appearance, the Exmoor Horn differs markedly from other hill breeds. It is a compact, chubby sheep, with a rounded conformation. It has a white face with a medium forelock and both sexes are

Exmoor Horn ram

horned. It is found mainly in its native area. Mature ewes weigh 52 kg and on the lowlands average 150% lambing. Draft ewes are crossed with suitable rams to produce crossbred breeding ewes.

Crossed with the Devon Longwool, the Exmoor Horn gave rise to the Devon Closewool, which is found in North Devon. It is a medium-sized, white-faced, polled sheep. Adult ewes weigh 61 kg and produce about 3 kg of wool with a staple length of 10 cm and a Bradford Count of 46–50. It is used as a grassland breed.

Finnish Landrace

Sheep in Finland are kept in very small flocks of 10 to 20 ewes and all belong to one breed, the Finnish Landrace. The outstanding feature of these sheep is their high prolificacy and they are selected mainly for this factor. The climate in Finland combines a hard, but dry, winter with a relatively low rainfall and continental summer. In addition the sheep are housed in the winter, so that they are not a hardy breed. Their fleece does not shed water so the Finnish Landrace is not suited to areas of high rainfall, consequently its place as a pure breed is limited.

A Breed Society was established in Finland in 1918, and sheep were exported to Great Britain in the 1960s. From there they spread to several other countries. This breed has been carefully assessed in these countries, and trials have confirmed its ability to achieve 250–300% lambing, and its long breeding season. The rams are aggressively sexual. At the same time butchers have a poor opinion of its conformation and carcase characteristics. Lamb mortality is high, milk yield is low, and the growth rate of the lambs is slow, so that the Finnish Landrace is less attractive as a prolific sheep than the Cambridge in Great Britain or the Romanov in France.

The Finnish Landrace is a small, pink-nosed, whitefaced breed, which is usually polled, although rams may be horned. Adult ewes weigh about 50 kg. The wool is soft and open. Despite the breed's deficiencies, it has made some contribution to the sheep industry of Great Britain, and lambs suffer no loss in quality provided that the Finnish Landrace contributes no more than $12\frac{1}{2}\%$ of their ancestry. The importation of Finnish Landrace sheep into Great Britain focused attention on the importance of prolificacy and provided an invaluable stimulus to sheep farmers to improve this factor.

The Romanov originated in Russia where the breed numbers about 700,000. They are kept in large flocks and are reasonably hardy. Their lambs are active at birth and show early sexual maturity. The fleece varies considerably in colour; the face is black,

Finnish Landrace ram

frequently with a white blaze. The tail is short, being only 13 cm long. The breed averages 250–300% lambing and the lambs are usually black at birth.

Hampshire Down

The foundation of the Hampshire Down breed dates back to the first Royal Show, in 1839, when William Humfrey of Cold Ash near Newbury in Berkshire saw and purchased a son of Webb's famous Southdown ram, Babraham. This ram, and others purchased subsequently, was used on the local varieties of sheep, the whitefaced Wiltshire Horn and the mainly specklefaced Berkshire Knot. From the mixture arose the Hampshire Down, which originally combined the adaptability to the short chalkland swards of the native breeds with the meat-producing ability of the

Southdown. The new breed was firmly established by the early 1840s and was superseding its parent breeds.

Classes for the Hampshire Down were started at the Royal Show in 1857 and the Breed Society was formed in 1889. In 1913 there were 485 registered flocks in Great Britain, while today there are only 65 registered flocks containing about 4000 breeding ewes. The breed increased its numbers in the USA after 1910 and is now a popular meat sire, especially under range conditions. It has also been exported to many other countries and can produce good lambs out of low grade native ewes.

The face is dark brown, almost black, and is well covered with wool over the poll and forehead. It is polled, but its origins are betrayed by a tendency to a Roman nose and occasional horn 'snags'. Mature ewes weigh about 80 kg, have a relatively long breeding season, and achieve an average 135% lambing. Fleece weights average 2·5 kg, with a staple length of 7 cm and a Bradford Count of 56–58.

Compared with other Down breeds the Hampshire produces crossbred progeny which are short in the leg but also short in the body. It transmits a lower growth rate than the Suffolk, but its progeny come to maturity more quickly. Hampshire rams on performance test achieve a daily liveweight gain of 320 grams.

Herdwick

Although there is a theory that the original Herdwick sheep swam ashore from the wreck of the Armada, the foundation of the breed can be traced from a mixture of the horned, blackfaced sheep of the Pennines and the Scandinavian tan-faced sheep that were introduced by the Vikings. The word 'herdwick' is an old north-country name for a sheep pasture.

(*above*) Hampshire Down ram

(*below*) Herdwick ram

At birth Herdwick lambs have a black face and legs, and a considerable proportion of black on the body. With increasing age the colour changes gradually so that in adult animals the face and wool is almost white. The wool is strong and coarse, with a staple length of 13 cm and a Bradford Count of 28–32. An average fleece weighs about 1·75 kg and is used mainly as carpet wool. Rams in particular have a strong mane or ruffle of wool round the neck. Rams are horned but the ewes are polled.

Adult sheep are short-legged and stockily built with strong bone. The Herdwick probably is the hardiest of British breeds of sheep, and has evolved in the harsh environs of the mountains of Cumbria. In common with other mountain breeds it has a strong hefting instinct which enables it to remain near the place of its birth, even on unenclosed moors.

The Herdwick has been restricted mainly to the Lake District of Cumbria, and it has been losing ground recently to the Swaledale, although new Herdwick flocks have been established in south-west England. It fits into the pattern of the stratification of the sheep industry in which breeds with differing characteristics are crossed to combine their qualities. It is used as the dam breed of crossbred breeding ewes, albeit in a relatively minor role.

Île de France

In 1833 Professor Yvart of the École Vétérinaire d'Alfort in Paris was breeding the French strain of Merino sheep, the Rambouillet. In that year he imported Dishley Leicester rams to cross with his sheep in order to obtain 'the Dishley in the wool of a Merino'. The result of this cross became known

Île de France ewe

eventually as the Dishley-Merino, but it was not until 1922 that a Breed Society was established and the name of Île de France was adopted. At that time the breed numbered about 1,500,000, or 15% of the French sheep population. Since that time the numbers of Île de France have declined so that in the late 1960s there were about 600,000 sheep representing 7% of the national flock. Nevertheless it is still one of the most important French breeds of sheep, and has been exported to most European countries, and to South Africa and South America.

The Île de France has inherited many characteristics from its Merino ancestors. It has a long breeding season with the ability to lamb in the autumn. It achieves on average 135% lambing, but this varies from over 140% with spring lambing to under 130% in the autumn. The Île de France has a white face and a pink nose, with a covering of wool over the poll. The wool is of high quality and a mature ewe produces a fleece weighing about 4 kg, with a staple length of 13 cm. An adult ewe weighs 70–75 kg.

The main purpose of the breed is as a meat sire for the production of high quality crossbred lambs. It is reputed to possess outstanding carcase characteristics, but these claims have not been confirmed entirely by comparative breed trials. These have indicated that the growth rate of the Île de France is lower than that of the Dorset Down. Crossbred progeny of Île de France rams have a relatively long carcase, but there is a tendency for too much weight to be carried in the forequarters which have fairly heavy bones.

Jacob

Multihorned sheep are found in many countries, but the islands around the coast of Scotland have been the home of several breeds of this type. Some, such as the white St Rona's Hill sheep, are now extinct, but three breeds can still be found: the Jacob, St Kilda and Manx Loghtan.

The Manx Loghtan is a very rare breed with little more than 100 breeding ewes. As its name implies, it is found mainly on the Isle of Man and the word 'Loghtan' describes its colour, derived from the Manx words 'Lugh' (mouse) and 'Dhoan' (brown). The St Kilda has spread throughout Great Britain but still numbers less than 1000 sheep. It is black in colour. Adult ewes weigh about 38·5 kg.

Until recently the Jacob also was considered a rare breed, but in 1969 a Breed Society was formed and there are now more than 350 members owning more than 3000 registered animals. The origin of the Jacob

Jacob ram

is obscure. It may be a member of the Hebridean group of multihorned sheep, or it may be related to the multihorned sheep of the Mediterranean, some of which were taken to North America by the Spanish Conquistadores and are now herded by the Navajo Indians.

Jacob sheep vary in type from the smaller parkland animals which weigh 42·5 kg to the larger improved sheep which weigh 60 kg and achieve a much higher level of production. They have usually two or four horns, but occasionally polled or six-horned specimens are found. The fleece has distinctive, large, black patches on a white background, and the wool is in great demand for home-spinning into undyed materials. The Jacob usually has a white blaze down its face and is a most attractive sheep.

Kerry Hill ram

Kerry Hill

In the 18th century the Eppynt Mountains in Wales were inhabited by a specklefaced variety of sheep of good body conformation. It was from this base that the Kerry Hill breed emerged. As with many breeds a small amount of Leicester blood may have been introduced, but the modern Kerry Hill is more probably an amalgamation of several Welsh Border types, with a dash of Down blood. Up to the middle of the 19th century there was very little uniformity within the breed. It was simply a rather bigger and stronger type of mountain sheep and bore some resemblance to the Clun Forest. In fact, at this time, draft ewes of both breeds were sold under the same banner, and were transferred to lowland farms where they were crossed with Leicester rams for fat lamb production.

Systematic improvement of the Kerry Hill began about 1850. The Breed Society was formed in 1892 and the first volume of the Flock Book was published in 1899. This breed is now kept mainly in self-contained flocks or is crossed with Down rams for fat lamb production. Small numbers have been exported to North America, South Africa and Australasia.

Traditionally the Kerry Hill has a black and white speckled face with sharply defined colouring. The nose is black, but modern fashion demands that other black markings on the legs and around the eyes are kept to a minimum. Ewes are polled, but rams on occasion carry small horns which are considered objectionable. The head is well woolled over the poll and on the cheeks, with a tuft of wool on the forehead.

Mature ewes weigh about 55 kg, achieve 157% lambing, and produce 2·5 kg of soft wool which has a wide range of uses. The staple length is 10 cm and the Bradford Count is 48–56.

Leicester Longwool

The history of the English Leicester is linked inevitably with Robert Bakewell, the master breeder who is acclaimed as the creator of modern methods of animal breeding. Before Bakewell, the Leicester was a long-legged, slow maturing, coarse animal, kept almost entirely for its yield of wool. Bakewell started his work around 1760 and his objectives were to produce 'a sheep possessed of the most perfect symmetry, with the greatest aptitude to fatten, and rather smaller in size'. He used the old Leicester sheep as his base and introduced Lincoln and Hereford (Ryeland) blood. From the prototype produced by this cross-breeding he isolated a quick maturing, fatter type, and fixed it

Leicester Longwool ram

by inbreeding. This process lasted about 20 years and became the accepted method in other breeds for establishing a standard type.

In 1790 Bakewell, who lived at Dishley Grange in Leicestershire, formed the Dishley Society. Only members of this Society were allowed to own the improved Leicester sheep and they were bound by rigid rules. Although Bakewell's methods were successful, his objectives must be questioned, and his sheep were strongly criticized by some of his contemporaries. They had lost prolificacy and milking ability; too much weight lay in the forequarters and the meat was of low quality; wool yield was reduced; and finally the large amount of fat in the carcase proved unacceptable.

The Leicester never became a dominant breed,

but its influence was felt throughout the sheep industry, where it was used to transmit its early maturity and finer bone to all other longwool breeds and to some middlewool breeds. It helped to create several new breeds including the Wensleydale (cross Teeswater), Lleyn (cross Welsh Mountain), Border Leicester (cross Cheviot) and Île de France (cross Rambouillet). It played a part in the development of the modern Wiltshire Horn, and abroad it contributed to the East Friesland, Texel and several French breeds.

The modern Leicester is a large, polled, whitefaced sheep. The Breed Society was formed in 1893, and since that time the size of the sheep and weight of the fleece has been improved. The average fleece weight is 5·5 kg of curly lustrous wool with a Bradford Count of 40–46. The breed is at present very limited in numbers with only about 250 registered breeding ewes. It is centred mainly on the Yorkshire Wolds, and its main purpose is the production of heavyweight lambs fattened off roots at eight to eleven months of age.

Lincoln Longwool

The Lincoln Longwool probably possesses the longest history of any of our sheep breeds. It originated in Lincolnshire as a big, slow maturing breed with poor mutton quality, but it produced the heaviest fleece of valuable long wool of any breed. At that time it was exceeded in size only by the old Teeswater. The Lincoln was used by Bakewell to produce the Dishley Leicester, which was then used to improve its parent breed. The Lincoln has been exported widely, especially to Argentina, Eastern Europe and Australasia. It proved particularly effective when crossed with the Merino, and both the Corriedale and Polwarth breeds were produced in this way. During the first three decades of the present century more than 70,000 Lincoln sheep were exported.

The Lincoln claims to be the biggest breed of sheep, and it produces 6·5–7 kg of lustrous longwool. The staple length is 35 cm with a Bradford Count of 38–44. It is polled, with a white face and a heavy forelock falling over the eyes. A record 81 cm staple length was produced by a yearling ewe, while a ram produced a fleece weighing 21 kg. There are now fewer than 1000 registered breeding ewes in about 20 flocks in Great Britain, although large numbers can still be found abroad.

At present the Lincoln is being used to sire heavyweight lambs but, although the carcases are lean, they tend to have relatively heavy bone with too much weight in the forequarters and a small eye muscle. It is likely that, as with the original type, wool will continue to be the most important product of the breed. Lincoln wool is still in demand for specialist trade such as lustre yarns; in particular the skins are ideal for the manufacture of sheepskin rugs.

Lleyn

In the various steps of development of Welsh sheep, the Lleyn represents a relatively advanced stage. Just as the Welsh Mountain was an improved type of the old tan-faced sheep, so in turn it was improved further by crosses from other breeds. In north-west Wales, Leicester and possibly Roscommon, rams were introduced from the mid-18th century onwards to sire the local ewes. The result of this cross was the Lleyn, which owes about three-eighths of its ancestry to the Welsh Mountain and five-eighths to the Leicester. It is claimed that Border Leicester rams may

(*above*) Lincoln Longwool ewe

(*below*) Lleyn ewe

have been used in the early 19th century, but there is no evidence to support this. The Lleyn is a very localized breed, being confined mainly to its original home of Gwynedd (formerly Caernarvonshire and Anglesey). In recent years interest in the breed has increased and there are now about 1500 breeding ewes found as far afield as the Welsh Borders and Devon.

The Lleyn is a small, polled sheep with a white face. Mature ewes weigh about 52 kg, although there is a good deal of variation. They are comparable to the Clun Forest for production characteristics such as prolificacy and milk yield. The Lleyn averages about 155% lambing, while single male lambs achieve a daily liveweight gain to 56 days of 335 grams and twin male lambs 240 grams. Individual sheep have outstanding performance records, including a ewe that produced 18 lambs in five crops. This ewe was used in the formation of the Cambridge breed which owes one-third of its ancestry to the Lleyn.

Merino

More than 2000 years ago, in the Near East, the local sheep were being transformed into a fine-wool type by deliberate breeding policies. Later the breed spread westwards, assisted by the Phoenicians and the Romans, and reached the Iberian Peninsula. The Moors developed the wool industry in Spain and the sheep, known as the Merino, became famous for the fineness of their wool. Until the end of the 18th century the Merino was guarded jealously by the rulers of Spain, but in spite of this it had spread to many parts of Europe and even to North America by AD 1810. France in particular imported large numbers

Merino ram

of Merino sheep and eventually established them as a separate breed, the Rambouillet.

King George III established a famous flock in England, where the Merino was at the height of its importance from 1810 to 1824, but thereafter its influence was negligible. It was reintroduced into England more recently, but once again it was not able to tolerate the wet climate. In North America various types were developed with the small Vermont strain at one extreme, having a heavily wrinkled skin to produce higher yields of wool. This type did not persist. In contrast, the Delaine Merino is a bigger sheep with mature ewes weighing 56 kg. It has an improved carcase and a higher yield of clean wool, owing to a lower proportion of grease in the fleece. The Rambouillet is also important in North America.

It was in Australia that the Merino attained the

height of its importance. The Tasmanian Merino is unsurpassed for the fineness of its wool and it carries five times more wool fibres per square centimetre of its skin than any other unrelated breed. No breed of sheep has contributed so much as the Merino to the development of the international sheep industry. The face is white and the flesh of the muzzle is pink. Rams are horned but ewes are hornless.

The Merino has contributed also to the creation of many other important breeds, either directly or through its sub-variety, the Rambouillet. Crossed mainly with longwool breeds, particularly the Lincoln, it has given rise to the Corriedale (New Zealand), the Columbia, Targhee, Panama, Montadale and Romedale (all USA), and the Île de France (France), while in England its influence can be seen in the Dorset Horn and the Whitefaced Woodland.

Portland

The Portland is one of the old tan-faced breeds that were the original inhabitants of the British Isles, and thus closely related to the Welsh Mountain and the Exmoor Horn. Together with Merino-type sheep, it formed the basis of the modern Dorset Horn. This acquired its pink nose and out-of-season lambing characteristic from the Merino, but in the process the Portland itself probably received a slight infusion of Merino blood, which added an extra spiral to the horns of the rams. This also accounts for the reputed ability of the Portland to lamb out of season. In the past the breed was considered to be prolific and to produce meat of great flavour and delicacy.

The majority of Portland ewes at present are lambed in January or February, and average little more than 100% lambing. This low level of prolific-

Portland ewe

acy may be due to inbreeding or to lambing early in the year. Only one flock successfully lambs more than once per year, and it is possible that the Portland has reverted to the tan-faced type. There is, however, no doubt about the quality of the meat. The Portland produces a lightweight carcase of fine-grained meat of excellent flavour.

It is a small sheep weighing about 33 kg; both sexes are horned. The face and legs are brown or tan, and when the lambs are born their wool is reddish-brown, changing to grey or white during the first year. The wool of the Portland is close and fine, but carries red kemp fibres in the britch, which indicates its relationship to the primitive tan-faced breeds. It is at present very limited in numbers, and only three separate bloodlines now exist, although most breeders are actively increasing the size of their flock.

Radnor

A good deal of confusion has arisen in the past by attempting to define the origin of various Welsh breeds in an area of the same name. Thus the Clun did not originate on the hills of the Clun Forest, while the Kerry Hill can be traced back to the Eppynt Mountain sheep. In the same way the Radnor emerged from the Kerry Hill in mid-Wales.

As late as 1920 the Radnor showed little sign of improvement, and it can be regarded as the most typical modern representative of the old Welsh tan-faced sheep. There was an attempt to form a Breed Society in 1926, but this was not established successfully until 1951.

The Radnor has a tan face which is free of wool. The nose is aquiline, especially in the rams. The ewes are polled, but the rams are usually horned. The horns, which are small, are closely set at the base and curve backwards close to the head. The Radnor is an intermediate type between the hill breeds and the grassland breeds such as the Clun and the Kerry Hill. It produces about 2 kg of rather kempy wool which is used in the production of Welsh woollen fabrics. The staple length is 13 cm and the Bradford Count is 50–58. Mature ewes weigh about 53 kg and achieve 145% lambing when brought to the lowlands.

Romney

The Romney sheep is probably the last typical representative of the original longwool sheep of Great Britain. It has evolved relatively undisturbed in that bleak and isolated area in south-east England known as Romney Marsh, where it has been exposed to cold sea mists and gales from the English Channel. The Romney is mentioned in the medieval wool lists and

(*above*) Radnor ram

(*below*) Romney Marsh ram

resembles the Cotswold at any early stage of its development. Low, writing in 1842, thought that the breed had 'coarse heads ... narrow chests, flat sides and ... fell well short in weight of the heavy-wooled sheep of the eastern counties ...'. There was some interchange of breeding stock between Kent and Flanders, but it was the cross with Leicester sheep in the early nineteenth century that deepened and rounded the body, refined the bone, and hastened maturity. In the present century Corriedale blood was introduced to improve the quality of the wool. The Breed Society was formed in 1894.

The Romney has been exported widely and has made a valuable contribution to the sheep industry in many countries. In particular a high proportion of the commercial ewes in New Zealand are derived from the Romney. In Great Britain it is crossed with the Southdown to produce a lamb for which there is a specialist demand in south-east England.

The wool produced by the Romney is classed as demi-lustre, is denser and finer than that of other longwools, and has a Bradford Count of 48–52. The average fleece weight is 4 kg, with a staple length of 15 cm. Mature ewes weigh 65–70 kg and achieve about 125% lambing. Romney sheep are polled, with a white face and a thick forelock. At present there are about 60 registered flocks containing about 15,000 breeding ewes.

Although Romney Marsh is cold and damp in winter, the pasture is lush and strong in the summer so that the sheep can be densely stocked. It is claimed that the Romney is resistant to diseases such as foot-rot and liver fluke, but this has not been proved by independent trials. The Romney is now crossed, usually with North Country Cheviot rams, to produce the Romney Halfbred. This is a relatively low-performance ewe that does not compare with more popular crossbred ewes such as the Mule or the Scotch Halfbred for either prolificacy or profitability.

Ryeland

The early Hereford Sheep enjoyed a widespread reputation for the quality and fineness of their wool, which was dubbed 'Lemster ore', and which was pre-eminent among the short wools of Britain until late in the 18th century. It was at this time that the name of Ryeland was first used, but in the 1790s the breed was 'improved' by the use of 'Down' rams which increased the weight of the fleece and the carcase, but coarsened both the wool fibres and the grain of the meat. At a later stage Leicester rams, and finally Southdown rams, were used so that the modern Ryeland bears little resemblance to the original upland Hereford sheep.

The average fleece weight of the Ryeland is 3 kg.

Ryeland ram

The staple length is 9 cm and the Bradford Count 56–58. The face is white with a well-woolled poll, forehead and cheeks. The breed performs a function similar to that of the Southdown, producing high quality, lightweight lambs. It is a minority breed in Great Britain.

Scottish Blackface

In common with the other important blackfaced, horned, mountain breeds, the Scottish Blackface traces its origin to the Pennine hills. From there, during the 17th and 18th centuries, it spread throughout Scotland displacing the old tan-faced sheep. At that stage it was known as the Linton or 'short sheep', as opposed to the Cheviot, which was the 'long sheep'. The Cheviot was also surging northwards through Scotland, and it is likely that the Blackface was influenced to some degree by introductions of both tan-faced and Cheviot blood. During its development, distinct types emerged of which the most important were the Lanark and the lighter fleeced Newton Stewart. The breed enjoyed the height of its fame towards the end of the 19th century. Thereafter the growing practice of housing ram lambs on systems of artificial feeding caused some loss of hardiness, but the Blackface still remains the most numerous mountain breed in the British Isles with more than 2,000,000 breeding ewes. Apart from its virtual monopoly of the Scottish Highlands, it has been established in Ireland and on Dartmoor, and has been exported to North America and France.

The Scottish Blackface is heavily horned in both sexes, and the rams' horns are especially imposing with a double spiral. The face and legs are mottled black and white, and are clear of wool. The face is

Scottish Blackface ram

strong with a pronounced Roman nose. This is a hardier breed than the Cheviot, but inferior to the Swaledale in this respect. Swaledale rams have been used in Blackface flocks to impart greater length of leg and improved milking ability, but the crossbred offspring also have appreciable hybrid vigour.

Under lowland conditions Scottish Blackface ewes weigh about 59 kg and average about 155% lambing. The wool is exported in considerable quantities, particularly to Italy where it is valued highly for filling mattresses. An average fleece weighs about 2·5 kg, and much of the wool is used for carpet manufacture. It is long and coarse textured, with a staple length of 30–33 cm and a Bradford Count of 28–32.

Ewes spend their first three productive years in hefted mountain flocks, after which they are sold as regular draft ewes. They are then crossed either with a Down ram or more commonly with a Border Leicester ram. The result of the latter cross is the Greyface, a popular crossbred breeding ewe in the lowlands.

Shetland

The early stages in the improvement of sheep can be traced through Scotland's relatively small number of native breeds. One line of development is illustrated in progression by the Soay, the North Ronaldsay and the Shetland, showing increasing size and the change of the colour of the wool from brown to mainly white.

Shetland sheep are noted primarily for the fineness and softness of their wool, which forms the basis of the well-known cottage industry of the Shetland Islands. The handmade lace shawls and scarves, and the Fair Isle patterned sweaters have achieved international fame. Although white is the predominant colour, there are several other natural shades of wool,

Shetland ewe

such as moorit (red), black, sheila (grey) and brown, and multi-coloured knitwear can be produced from undyed wool. Fleeces weigh on average rather more than 1 kg. The staple length is 10 cm and the wool, which has a Bradford Count of 54–60, is shed during the summer if not previously plucked or shorn. The softest wool is produced on the poorest, dampest moor pastures, and traditionally the wool from the sheep in North Maven is considered to be the finest.

The true Shetland sheep is small, with pointed erect ears and an alert appearance. It has a short tail, which is broad at the base, known as a 'fluke tail'. This breed is closely related to Scandinavian sheep. Rams of other breeds have been introduced at various times, but the severe conditions under which the sheep must survive and produce suit only those animals which possess the native characteristics of the Shetland breed to the greatest degree. Ewes are usually polled, but the rams carry light horns. A Breed Society was formed in 1926.

Shropshire

The Shropshire Down arose from a complex variety of breeds that existed in the West Midlands and Welsh Borders, but the Longmynd, a small black-faced horned sheep, and the Morfe Common, a horned specklefaced fine wool sheep, were the main contributors to its formation. In the early stages Southdown and maybe Leicester rams were introduced, so that it was not until the middle of the 19th century that the breed emerged as a fixed type, and it was given its own classes at the Royal Show in 1859. In 1882 the Breed Society was founded, and in 1883 published the first Flock Book.

This breed was at the height of its popularity in the

early part of the 20th century, and in 1906 over 2000 registered animals were exported. It was bred mainly for the export market, especially the USA, but American breeders demanded a smaller type of sheep with a heavily woolled head. When this market ceased fairly abruptly about 1930, the breed declined rapidly. Now there are only 14 registered flocks, containing fewer than 450 breeding ewes, concentrated mainly on the Midlands and Welsh Borders. But the Breed Society is active and the breed is increasing in popularity once again.

The Shropshire now has a naturally clean soft black face with a covering of wool on the poll. An average fleece weighs 3·5–4 kg, with a staple length of 10 cm and a Bradford Count of 50–58. The growth rate of the crossbred progeny of Shropshire rams is intermediate between the progeny of the Suffolk and the Southdown, but it is probably the hardiest and the most prolific of the Down breeds.

It has never found the favour in Great Britain that might be expected, perhaps because it falls between the two stools of quality and quantity, but its general purpose ability enabled it to achieve great popularity abroad. It has been exported in large numbers to North America, East Africa and Australasia, where it was valued highly for its ability to produce quality lambs out of Merino-type ewes.

Soay

Of all the domesticated and feral breeds the Soay bears the greatest resemblance to wild sheep, particularly the Moufflon of Corsica. The breed has survived mainly in the Outer Hebrides on the islands of Soay and Hirta, which are fully exposed to the Atlantic storms. Soay is a Norse word which means

(*above*) Shropshire ram

(*below*) Hirta Soay ewe

sheep island. About 1500 feral Soay sheep exist on these islands, and at various times groups of sheep have been transferred to the mainland, where they have been kept for aesthetic purposes in parkland.

The Soay is a small, primitive, short-tailed sheep. Ewes on the mainland weigh 26 kg, and may be either horned or polled. Rams stand about 55 cm at the withers and weigh about 40 kg. They develop a dark, hairy mane and have heavy horns which grow in a single plane with no spiral. The back and flanks are either dark brown or fawn in a ratio of 3:1, but the belly and chest are lighter coloured and there is a light patch on the rump. Occasionally black, white and piebald animals are seen. The fleece contains both wool and hair, and in some cases the coarser fibres predominate and project beyond the rest of the coat to give a hairy fleece. The variety of both colour and fleece type illustrates how it was possible to develop modern breeds of sheep from the Soay. The staple length is seldom greater than 5 cm and the Bradford Count is 44–50 for the finer wool. The fleece is normally shed and weighs on average 0·7 kg.

Although it has been assumed generally that the Soay has no commercial significance, recent research has shown that it can play a valuable role in low cost farming with minimal supervision. The meat has a distinctive flavour and finds a ready market. Although some flocks of Soay ewes average less than 100% lambing, with better management this can be raised to 150%. Purebred lambs achieve a daily liveweight gain of only 115 grams, but this growth rate can be doubled by crossing Soay ewes with Devon rams.

The Soay is a vital link between wild and domesticated sheep. While the domesticated flocks on the mainland may be developed for commercial characteristics, the feral flocks will remain as a reservoir of unique genetic material. Small breeding groups of Soay sheep have been exported to the mainland of Europe and to North America.

Southdown

About 20 years after Bakewell began the experiments which resulted in the Dishley Leicester, another breeder embarked upon a programme which was to exert an even greater influence upon the sheep industry of Great Britain and the world. He was John Ellman of Glynde in Sussex, and he took as his raw material the local heath breed of the Downs, which was a fine-boned, leggy sheep, with light forequarters and a speckled face. Ellman has left no records to show how he transformed this sheep into the blocky, compact animal known as the Southdown. Ellman's work was continued by Jones Webb of Babraham in Cambridgeshire, and the fame of the breed spread

Southdown ram

rapidly. The Breed Society was formed in 1892.

This variety was exported in the early 19th century to the mainland of Europe and to North America, and until recently it was the dominant meat ram in New Zealand. In Great Britain it is used mainly to cross with Romney ewes to meet a specialist demand in south-east England. Today the fashion among breeders is to select for shorter legs and a lighter coloured face, and the darker, bigger Southdown in France is probably more truly representative of the old type. French breeders have retained more size, prolificacy and thriftiness in their Southdown sheep, whose progeny must be able to fatten on the thin mountain pastures of the Massif Central.

The modern Southdown is the smallest of the Down breeds. Mature ewes weigh about 60 kg, and the crossbred progeny grow relatively slowly but fatten more easily. Like the Improved Leicester, the importance of the Southdown is found less in its own qualities than in the part it played in creating new breeds such as Suffolk, Hampshire Down, Dorset Down, Shropshire, Ryeland, and some French breeds. It has a mouse-coloured face, the upper part and the ears being covered with wool. It produces about 2·75 kg of wool, which is of high quality, with a Bradford Count of 56–60. The staple length is 6 cm. In Great Britain the Southdown is a minority breed with only about 1000 breeding ewes, but it is still widespread abroad, and the breed probably numbers 500,000 sheep in France alone.

Suffolk

In 1786 Arthur Young, later Secretary of the Board of Agriculture, used a Southdown ram on a group of Norfolk Horn ewes, and by 1791 had 350 crossbred

Suffolk ram

ewes. Almost 200 years later the Norfolk Horn is extinct and the Southdown is a minority breed in its country of origin, but the inbred cross between these two breeds, now known as the Suffolk, is one of the most popular international breeds of sheep. It has been recognized as a pure breed since 1810, and the Breed Society was formed in 1886. The first Flock Book was published in 1887 with details of 46 flocks, while today there are more than 900 registered flocks in Great Britain alone, containing about 35,000 breeding females.

The Suffolk was developed originally as a breed suited to the arable systems of East Anglia, but is now found in a wide range of conditions. It was introduced into North America later than the other Down breeds, and achieved popularity there only after the 1920s, but it is now a leading breed especially in the range country, from the mid-West to the foothills of the Rocky Mountains. The Suffolk inherited from the Norfolk Horn its black face and legs, a degree of hardiness, prolificacy, a lean carcase and a tendency to grow horn buds, while the Southdown cross introduced the characteristics of docility, quick maturity and a better carcase. It has a long body and short close fleece. The average fleece weight is 3 kg, with a staple length of 6·5 cm and a Bradford Count of 54–58. It is polled, and the relatively large ears are carried horizontally. Mature ewes weigh about 80 kg.

The breed is more prolific than most other Down breeds and achieves an average of about 150% lambing, which is comparable to the Shropshire. The crossbred progeny of the Suffolk achieve better growth rates than other Down breeds, apart from the Oxford Down, but tend to mature later and are best suited to the production of heavyweight lambs in the autumn. Purebred Suffolk ewes have an extended breeding season and are able to lamb earlier than most other breeds, to produce finished lamb for a high price market.

Swaledale ram

Swaledale

The north of England is the ancestral home of black-faced, horned breeds. The Swaledale, a breed of increasing importance in this group, originated in a pocket of the Pennines between Swaledale Head and Stainmore in North Yorkshire. From here it has spread to all the northern counties of England, has invaded the territory of the Scottish Blackface north of the Border, and is found in the south of England—particularly in Devon, Sussex and East Anglia. It has been exported recently to France. A Breed Society was formed in 1919.

The Swaledale has evolved in a harsh, exposed environment, and is probably Britain's hardiest breed apart from the Herdwick. Its fleece consists of an outer layer ('coat') of long coarse fibres which shed the rain, and a dense inner layer ('waistcoat') of finer wool which prevents the cold winds penetrating to the skin. The breed maintains its hardiness and thriftiness because rams are bred in commercial flocks managed on a normal hill farming system.

Both sexes of this breed are horned, and the horns of rams have an impressive double spiral. The face is black with a white or grey nose, and the legs are mottled. A mature ewe weighs 53 kg and produces 1·8 kg of wool, which is used for the manufacture of tweeds or for tough carpet yarns. The staple length is 25 cm and the Bradford Count is 28–40. It is a prolific breed and achieves 162% lambing when brought down to the lowlands, and produces a good yield of milk. It is the dam breed of the Mule and the Masham, which are sired by Bluefaced Leicester and Teeswater rams respectively, and which are Britain's most productive crossbred ewes.

The Swaledale now numbers 500,000 ewes, of

which more than 50% are mated to Swaledale rams. Other closely related Pennine breeds include the Dalesbred, Lonk and Rough Fell, which together number over 300,000 ewes.

Teeswater

The Teeswater 'Mug' was the biggest of the ancient longwool breeds, surpassing even the Lincoln to which it was closely related. It was a long-bodied, coarse, slow-feeding animal, but the early breeders created a heavy, fine-boned carcase by selecting for a broad back. The Teeswater at that time also produced the heaviest fleece, but the size of the sheep and the weight of the fleece both declined towards the end of the 18th century with the introduction of Dishley Leicester blood. The Teeswater was noted at an early stage for its prolificacy. Twins were common, and quadruplets and quintuplets were encountered. One ewe is recorded as producing 20 lambs in six years, while in 1802 a group of 24 Teeswater ewes produced 70 lambs.

The breed fell into disuse during the 19th century, but remnants of the old type persisted in Upper Teesdale on the boundary between Yorkshire and Durham, and after the Second World War the breeders in that area re-established the breed; a Breed Society was formed in 1949 incorporating 185 flocks.

The modern Teeswater is smaller than the old 'Mug' of the 18th century. It is a white or grey-faced, polled breed. The average fleece weight is 6 kg, with a staple length of 30 cm and a Bradford Count of 40–48. The production of crossing rams is the prime objective of Teeswater breeders, who progeny test their rams on Swaledale ewes to produce the Masham. The Masham is one of the most prolific crossbred ewes, but trials have shown also that the progeny of Teeswater rams out of hill ewes achieve higher growth rates than the comparable progeny of the Border Leicester or the Colbred.

Crossed with the Dishley Leicester, the old Teeswater gave rise to a new breed, the Wensleydale,

Teeswater ram

which by a twist of fate was superseded as a crossing ram breed by the new Teeswater. However, in recent years the Teeswater is being ousted again by yet another related breed, the Bluefaced Leicester, and fewer than 1000 breeding ewes now remain.

Texel

The Texel breed of sheep originated on the island of the same name off the coast of Holland. It was based on the native Marsh sheep, which was improved in the present century by crossing with imported English rams of the Leicester, Lincoln and Kent breeds. The new breed spread into neighbouring countries, especially France where it was introduced in 1933, and it has been exported from there recently into Great Britain.

There are now two distinct types of Texel sheep. The original Dutch type, which is related to the East Friesland breed, is both prolific and milky, but in France the Texel has been selected more for its meat-producing potential. It is increasing in popularity in France, which is now the main centre of the breed, and a Breed Society was formed in 1935. Texel sheep have a white face clear of wool, with large ears, and resemble a large strain of the Lleyn. This is a grassland breed which has evolved in a cold area of high rainfall. It experiences significant difficulties at lambing, and East Friesland crosses are being introduced to reduce this problem.

Although the original Dutch Texel was relatively prolific and achieved 175–195% lambing, the French type averages less than 160% lambing. An adult ewe weighs about 75 kg and produces a fleece of 4·5 kg with a staple length of 15–20 cm. The greatest interest in the breed now is directed to its potential as a meat ram to sire crossbred lambs of high carcase quality. It transmits to its progeny a growth rate that is between the Suffolk and the Hampshire Down. A higher proportion of the weight is carried in the forequarters, but there is a high percentage of lean meat in the carcase, with a low fat content, which makes it

Texel ram

acceptable for the modern trade. It has a particularly large eye muscle, and it is this quality which has excited the current interest in the breed.

Welsh Mountain

The sheep inhabiting the mountains of Wales are an excellent illustration of the development and improvement of the general type of British sheep. The original type was the small, tan-faced Rhiw and Cardy sheep, which became extinct within the last 20 years, and from this base have emerged several closely related breeds within the Welsh Mountain group. These include the Hardy, Aberystwyth, South, Black and Speckleface varieties. It is not unusual for 'Badger face' reversions to the old type to occur in many modern flocks.

The Aberystwyth Welsh Mountain is the most common variety. It is a small, hardy sheep with a white face and white legs, although red kempy hairs are found in the fleece and indicate the tan-faced ancestry of the breed. Rams are horned. Adult ewes weigh about 40 kg and are not prolific, averaging little more than 110% lambing even when brought onto better quality land. They produce about 1·25–1·5 kg of relatively soft and fine wool. The staple length is 6–9 cm and the Bradford Count is 36–50. The South Wales Mountain produces a heavier fleece with a higher proportion of kemp, while there is a specialist demand for the wool of the Black Welsh Mountain.

In total, Welsh Mountain sheep number about 2,000,000 breeding animals. They are found mainly in the mountains of Wales, the Hardy and Aberystwyth varieties in the north, the South and Speckleface varieties in the south, while the Black Welsh Mountain has spread more widely in mainly parkland locations.

The main purpose of the Welsh Mountain sheep is to utilize poor quality mountain grazing. Draft ewes are crossed, usually with a Border Leicester ram, to produce the Welsh Halfbred, which is one of the crossbred ewes employed by lowland flockmasters. The Welsh Mountain was used also in the creation of the Lleyn sheep.

Wensleydale Longwool

The Wensleydale breed of sheep originated in North Yorkshire as the result of crossing the old Teeswater longwool sheep with Leicester rams. One lamb born from this cross was the outstanding ram 'Blue Cap', who 'possessed a wonderfully broad masculine head of a deep blue colour', and who became the foundation sire of the breed. Blue Cap was born in 1839. The

(*above*) Welsh Mountain ram

(*below*) Wensleydale ram

breed type was established by about 1860, and the first volume of the Flock Book was published in 1890.

The Wensleydale is a large, longwool sheep. Shearling ewes average 77 kg liveweight and shearling rams average 100 kg, but mature rams frequently weigh more than 135 kg. The breed has a broad, level back, and possesses a superior conformation compared with most other longwools. The skin of the face and ears carries a distinctive, deep blue pigmentation, which may extend to the rest of the body and which causes a small proportion of the lambs to be born black. The wool is bright and lustrous and purled. It has a Bradford Count of 40–48, with a staple length of 30 cm, and an average fleece weight of 6·5 kg. The wool is used in the Yorkshire woollen

Whitefaced Woodland sheep

industry to produce a lustrous finish, or to blend with shorter-stapled wools to produce a strong yarn.

The Wensleydale is now found in only limited numbers, mainly in North Yorkshire, but with flocks in Cumbria, Lancashire, Lincolnshire and the West Indies. At present there are 22 flocks with a total of 231 breeding ewes, but only two flocks contain more than 30 ewes.

Demand for the breed is now increasing. Originally its main use was to cross with hill sheep to produce the Masham, one of the most prolific crossbred ewes in Britain. In this function it has been superseded by the new Teeswater, and it has now changed its role. It is at present used to sire heavyweight lambs in addition to its value as a wool-producing sheep. The Wensleydale has been used also to improve sheep in other countries and was one of the parent breeds of the Bleu du Maine, a leading breed in France.

Whitefaced Woodland

The Whitefaced Woodland is a member of the group of horned mountain breeds, but is unusual in various respects. It has a white face with pink or part pink nostrils. The rams of this breed are much larger than the ewes; these sheep possess a particularly strong, muscular tail. The Whitefaced Woodland is now concentrated mainly in that area of the Pennine hills where the three counties of West Yorkshire, South Yorkshire and Derbyshire meet, and it took its name from the valley, the Woodland Dale, that runs from Snake Pass to the Ladybower Reservoir. An alternative name for the breed is the Penistone, after the market town a few miles away in South Yorkshire where a sheep fair has been held since 1699.

The breed arose from a mixture of the native sheep of the Linton group with imported Merinos, together with a limited introduction of Cheviot blood. As might be expected, variations in type arose from such a mixture, and when the term Woodland was used first it referred to the tall, leggy type as opposed to the more compact Penistone. In recent decades these various types have been blended together and given the name of Whitefaced Woodland.

As recently as 1912 it was recorded that 'a peculiar feature of the ewes of this variety is that they come into use at any time—which is a recommendation as the lambs can be bred for the Christmas market'. The modern Whitefaced Woodland makes no claims on this point, and its main purpose is to cross with ewes of the blackfaced, horned mountain breeds to improve the size of the latter and to produce a speckle-faced ewe that shows considerable hybrid vigour. It is probably the largest of the mountain breeds, and rams weighing in excess of 130 kg have been recorded.

It is not so hardy as the Swaledale or the Scottish Blackface, and can be compared more closely with the Lonk and the Derbyshire Gritstone. Brought down to the lowlands, the ewes average better than 150% lambing. The breed has recently been introduced into new areas and can be found now in Scotland, Wales, Cumbria and the south of England.

Wiltshire Horn

The Wiltshire Horn is unique among improved breeds of British sheep. It does not produce wool, but instead its fleece is a thick matted covering, consisting mainly of hair, which peels as the sheep fattens. Originally the breed belonged to the whitefaced, horned group of sheep of south-west England, and still bears some resemblance to an old-fashioned, leggy Dorset Horn. Descriptions of the Wiltshire Horn in the late 18th century indicate that it was a fine middlewool, but that the covering was thin and sometimes the underparts were bare. Continuing selection has exaggerated this characteristic to produce the present hairy fleece.

By the end of the 18th century the Wiltshire Horn was being replaced by the Southdown in all parts of its native county except the Vale of the White Horse. This was the end of its era as a pure mutton-producing breed. It became extinct in Wiltshire about 1820, but had been established previously in the eastern and Midland counties. It survived in these areas but probably not as a pure breed owing to crossing with Leicester and Cotswold sheep. Around 1850, breeders

Wiltshire Horn ram

from North Wales became interested in the breed, and today the main centres are Northamptonshire and Anglesey.

The Wiltshire Horn is a distinctive, whitefaced sheep, and both sexes are horned. Because it does not grow wool it suffers less than other breeds from fly troubles. It is a robust breed, and the lambs are active at birth, while the loss of wool has been compensated by extra fleshing, especially over the back. Its main purpose at present is as a specialist meat sire for crossing with Welsh ewes.

CATTLE

Aberdeen Angus

In the early 19th century the black, horned, slow-maturing cattle in the county of Fife were crossed with more recently introduced red cattle of a variety of types, and the resulting mixture became part of the modern Aberdeen Angus breed. Black polled cattle in the north-east of Scotland are mentioned in ninth century records, and this native stock also contributed to the breed. The Aberdeen Angus was established as a type by Hugh Watson of Forfarshire who used inbreeding to fix the qualities of his cattle, but the Aberdeen Angus was popularized by William McCombie of Aberdeenshire, who exhibited abroad with great success. The Herd Book was established

Aberdeen Angus bull

in 1862 and included Galloway cattle until 1877. In 1879 the Aberdeen Angus Cattle Society was formed.

The coat colour is now uniformly black, although little more than 100 years ago the colours were extremely variable including yellow and brindle, with belted and line-back colour patterns. Red calves are occasionally produced and these have been used to establish the Red Angus breed. The Aberdeen Angus has a short, broad head and is always polled. The body is long and deep, and the legs are short and fine-boned. The breed is noted for its beef conformation and the quality of its meat.

It has enjoyed a considerable international reputation. The first effective importation to North America was made in 1878, and it has been used widely as a crossing bull to impart qualities of meat production to low grade native cattle in many countries. The bulls are popular because they colour mark their progeny and cause few calving difficulties. The Aberdeen Angus is an early maturing animal which will fatten on relatively low energy feed. Heifers calve at a young age and calves have a low birth weight. The growth rate of the calves is not high, but the finished animal gives a high carcase yield of well marbled meat. Bulls at 400 days of age weigh on average 400 kg and are 110 cm high at the withers.

Ayrshire

The main phase of improvement of the dairy cattle of south-west Scotland began in the second half of the 18th century. At this time Shorthorn or Dutch stock was introduced to cross with various local types, and from the mixture emerged the Ayrshire, the first of the modern specialist dairy breeds. It seems likely

Ayrshire cow

that the development of the breed was influenced also by some use of either Channel Island or Scandinavian cattle. Apart from its genetic quality, the Ayrshire has the added advantage of being close to the densely populated area of Glasgow, which provided a ready market for milk and dairy products. Thus the breeders had great incentive to improve the yield of their cattle and they accepted the challenge successfully.

The Ayrshire Cattle Society was formed in 1877 and the first Herd Book was published the following year. Official milk recording began in 1903. The Ayrshire has been exported to several countries, particularly those of North America and Scandinavia.

A mature Ayrshire cow weighs 460–500 kg and shows a typical dairy wedge-shaped conformation. The udder is symmetrical and the teats are well positioned but often rather small. The cattle are normally brown and white in colour, although a few are black and white. The horns are slender and grow outward and forward, and help to give the cattle an alert and stylish bearing.

Ayrshire cattle are hardy and thrifty, and are noted for their freedom from disease. They produce milk of high quality with a butterfat content of 3·86%. The fat is composed of small globules, which makes it suitable for the manufacture of cheese. The average yield of milk is 4384 kg in 305 days. The Ayrshire is not noted for meat production and this has been the main reason why it has lost ground to the British Friesian in recent decades.

Brahman

India is the home of Zebu cattle and each province has developed its own variety. The principal varieties that have been imported into the USA are the Guzerat, Nellore, Gir and Krishna, and the American Brahman Breeders' Association, which was established in 1924, is now an international organization. Large numbers of Indian cattle were sent to Brazil between 1870 and 1910, and from there found their way to Texas, where the name 'Brahman' was adopted.

All Brahman cattle are characterized by an abundance of loose skin, particularly on the throat and dewlap, with well-developed sweat pores which allow easy perspiration. They also produce an oily secretion from the sebaceous glands which repels insects. They have extended spinal processes over the neck and shoulder which are covered with muscular tissue to produce a marked hump. The head is long, and the horns usually curve upwards and sometimes are

Cotswold sheep in Gloucestershire parkland

PLATE 1

Wiltshire Horn sheep shedding fleece in June

(*above*)
Soay ewes and lambs

PLATE 2

(*right*)
Wensleydale ram in full fleece

(*left*)
White Shetland sheep

PLATE 3

(*top*) A typical English Longhorn bull
(*bottom*) A herd of Shetland cows grazing moorland in Scotland

PLATE 4

White Park bull and cow both 16 years of age

PLATE 5

An imported Chianina bull at the Great Yorkshire Show

PLATE 6

(*top*) Belted Galloway cattle
(*bottom*) A champion Cleveland Bay Horse

PLATE 7
Team of Shire Horses ploughing

Middle White pig

PLATE 8

'Iron Age' sow (Tamworth cross Wild Boar) with a litter sired by a Tamworth boar

Brahman bull

tilted backwards. The ears are large and drooping. The hair colour is most commonly steel-grey, but this varies from almost white to black, while red is usual in the Gir strain. The skin is pigmented and the colouring is darker on the forequarters and hindquarters than on the rest of the body. Mature bulls weigh about 860 kg in working condition, and mature cows weigh 545 kg.

The modern Brahman possesses many valuable characteristics. It is tolerant of high temperatures, and is resistant to insects, and diseases of hot climates such as eye cancer and a fever called anaplasmosis. These cattle are long lived, and in high temperatures achieve a higher level of production than European breeds. Under range conditions they are often wild and difficult to manage, but when handled correctly in enclosed areas they become very docile.

British White

The Vikings are credited with the introduction of polled cattle into Britain. Although such a generalization may oversimplify the situation, it is a fact that areas where Viking influence was greatest tend to be the home of polled breeds such as the Aberdeen Angus, Galloway, Suffolk Dun (Red Poll) and British White. The link between the British White and Scandinavia is particularly strong. In Sweden there is a mountain dairy breed, the Fjallras, which resembles very closely the British White, whose origin can thus be traced back through tribal migrations to Scythia. The Breed Society was founded in 1918 as the Park Cattle Society, which included also White Park cattle, and the first volume of the Herd Book was published in 1920. In 1946 the breeders of the polled cattle formed their own Society called the British White Cattle Society.

British White cow

At various times in the past breeders have used bulls of other breeds. In 1949 Fjallras bulls were imported from Sweden, while earlier, white Shorthorn bulls were introduced. Records show that in the past the British White was a good dairy animal, with individual cows yielding in excess of 5500 kg of milk per lactation. More recently White Galloway bulls have been used in grading up programmes, so that the emphasis has changed towards a dual-purpose type, and British White cows now yield on average only 3145 kg of milk per lactation with 3·74% butterfat. Cattle of this type have been exported to several countries including Canada, Kenya and Colombia.

The British White is a polled breed, and the cattle are at present rather variable in type. Generally they are short-legged, medium-sized animals with mature cows weighing about 450 kg and mature bulls about 600 kg. They are white in colour, with black or red ears, nose, feet and teats. The same colour pattern can be found in the Blanco Orejinegro cattle of Colombia, the Fjallras cattle of Sweden, and in individual cattle of the Texas Longhorn and Fulani breeds.

Charolais

The Charolais breed has been developed gradually over a long period of time in central France. It is descended from an indigenous breed of cream-coloured cattle which inhabited the Charolles district, and which originated from the same ancestral stock as the Simmental. Initially it was a triple-purpose type used for draught, milk and meat, but in the 18th century white Shorthorn bulls from England were used, and subsequent selection for growth rate and carcase quality has made the Charolais one of the most important specialist beef breeds in the world.

A Herd Book was opened in 1864, and since that time Charolais cattle have been exported to most parts of the world. They have proved particularly useful for crossing with Zebu cattle in North Africa, Oceania, South America and Mexico. From Mexico they reached Texas, where the Charolais cross Brahman was fixed as a new breed called the Charbray. Meanwhile the Charolais was exported to Britain in 1962, and from there it again spread to North America.

The hair is white or cream in colour, and is soft, and may be long when cattle stay out of doors during the winter. The head is short and broad, and the horns, which are medium-short, turn forwards and upwards. Charolais cattle are large and heavy, and bulls develop a strong muscular crest. The body is long, wide and deep, and extremely heavily muscled. Some animals, known as culards, have abnormally thick muscling of the hindquarters, and this frequently causes difficulties when the calves are born. The legs are short with heavy bones, and the hind legs are sometimes too straight at the hock.

Charolais cattle are poor producers of milk, although the cows provide sufficient to rear a calf, but they achieve higher growth rates than most other breeds. Bulls at 400 days of age weigh on average 580 kg and their progeny are noted for the high proportion of lean meat in the carcase.

Chianina

White cattle have been important in Italy since they were used for sacrificial purposes by the Romans, when the Empire was at its height. Today there are several white breeds including the Piedmont, Romag-

(*above*) Charolais bull
(*below*) Chianina bull

nola, Chianina, Marchigiana and the lyre-horned Maremma, and they are all descended from the old Grey Steppe cattle of south-eastern Europe. Until recently these breeds were used mainly as draught oxen, but the increasing use of tractors drew attention to their meat-producing potential. The Chianina is the most important breed, and its Herd Book was established as recently as 1956, but it has quickly assumed an international role.

The hair is porcelain white in colour, but the hooves, muzzle and points of the horns are black. The horns are short and curl forwards, and the head is long, with much loose skin around the throat. The dewlap also is well developed. The Chianina is one of the biggest breeds in the world. The legs are very long, so that mature bulls have a withers height of 170 cm and weigh on average more than 1200 kg. Cows stand 158 cm at the withers and weigh 850 kg. The body is long and cylindrical, and the withers are raised higher than the back, with a well-developed crest in the bulls.

As they were bred for the plough, Chianina cattle possess the muscularity and strong frame necessary to provide traction power. As work animals they were also very docile, and these characteristics are now being exploited for meat production. The skin is rather thin and pigmented, so that it is in countries with a hot climate that the Chianina is most likely to achieve an extremely high growth rate and high quality of meat.

Devon

The Devon is one of the old red breeds of Britain. It has been found in the south-west of England, where it has remained as a local breed for several centuries.

Devon cow

A population of red cattle was recorded in North Devon in the 11th century, and probably these animals were the origin of the modern Devon breed. They have been noted as draught oxen because, although rather small, they were agile and hard workers. Later some of the breed were found further afield, and in the late 18th century small herds were found in Northumberland, Warwickshire and Norfolk.

Devon cattle are known popularly as 'Red Rubies' because of their deep cherry red colour. The hair is fairly long and frequently curly, and the skin carries an orange pigment. The body is blocky with deep hindquarters, and the legs are short. The head is short and the horns of medium size. At 400 days of age bulls weigh 477 kg and have a withers height of 118 cm.

Devon cattle were formerly triple-purpose animals, and it is only in relatively recent times that the emphasis of selection has been for meat production.

Consequently it is still possible to find dual-purpose strains, and most Devon cows produce rich milk, containing more than 4% butterfat, in sufficient quantity to feed their calves extremely well.

Dexter

The origins of the Dexter breed of cattle are uncertain. It first became known in Ireland during the 18th century, and it may have arisen from a cross between Kerry and Devon cattle in order to produce a miniature breed suitable for smallholders. Alternatively the dwarf characteristic may have been imported from Africa, or have been established by mutation and selection within the Kerry breed. It is believed that the dwarf phenotype is shown by animals which can produce non-viable 'bulldog' calves. Dexter cattle were exported to England in 1882 and the English Breed Society was founded in 1900. Meanwhile the Irish Herd Book lapsed, but the Dexter has been established successfully in South Africa.

Mature Dexter cows weigh on average 300 kg, and bulls up to 400 kg. The legs are short and the bones of the feet and legs may be slightly deformed where the dwarf characteristic is strongly exhibited. The colour is generally black, although red cattle are also found and recently dun animals have appeared. Dexter cattle are dual-purpose. They are horned with a short, broad head, and the body is compact and deep in proportion to the size, with well-sprung ribs.

The main advantages of the Dexter are its ability to use poor quality grazing and its low feed requirement. It needs only 60% of the feed that larger cattle of the more popular dairy breeds consume. Thus it is a relatively efficient producer of both meat and milk.

Dexter cow

The average yield is 2295 kg of milk per lactation, with a butterfat content of 3·97%. Dexter cattle fatten easily and produce small joints of good quality meat. They are used most effectively either in small dairy herds on poorer land, or as suckling cows crossed with a beef bull and rearing two calves.

Dutch Belted

Dutch Belted cattle originated in Holland before the 17th century. They were kept by the nobility who bred animals to a particular colour pattern, in this case a white 'belt' which encircled the body between the shoulders and the hips. The Dutch developed this colour pattern in many species of livestock including rabbits, goats, poultry, swine and cattle. The belted cattle in Holland, which are called Lakenvelder, are now very rare. The name Lakenvelder is derived from 'Laken' (a sheet) and 'Veld' (a field) from the impression that the cattle have a sheet wrapped around them. When the colour pattern spread to England the same notion persisted in the name Sheeted Somerset.

Belted cattle were first exported to North America from Holland in 1838. These were followed by a second importation in 1840 by Mr P. T. Barnum, the well-known showman, who exhibited them before transferring them to his farm. The Dutch Belted Cattle Association of America was formed in 1886 but, although there were about 1500 registered animals by 1916, the breed has never achieved great popularity and is now relatively rare. Dutch Belted cows are excellent dairy animals averaging 4795 kg of milk per lactation with an average butterfat of 3·94%. They are found scattered throughout the eastern states of the USA, from Florida to New England, and seem capable of tolerating a wide range of climatic conditions.

Dutch Belted cattle at birth weigh about 35 kg, while mature cows weigh 545 kg and mature bulls 770 kg. They resemble the Ayrshire in type, having a definite dairy character, although the horns are shorter and lighter. Polled specimens are found, and these may result from the introduction of Galloway blood. The colour is black with the distinctive white 'belt' that gives the breed its name.

Friesian *(Holstein)*

Although cattle breeding was important in Holland as early as the 12th century, it is unlikely that the animals used were the ancestors of the Friesian, which is now the dominant breed in Holland. The old breed was reputed to be of great size and red in colour, but severe flooding from the Zuider Zee and outbreaks of cattle plague, especially in the 18th

(*above*) Dutch Belted cow

(*below*) Holstein-Friesian cow

century, reduced the numbers of this breed to such a low level that black-and-white cattle were imported from Jutland; these formed the main basis of the modern Dutch Friesian. The breeding methods of the Dutch farmers were very successful and their cattle were famous for their high yields of milk. They spread into Germany at an early stage, and were exported in large numbers to Great Britain as early as the 17th century. A Breed Society was formed there in 1909. They did not gain a firm foothold in France, where they are known as the FFPN, until after the Second World War, but they are now the most numerous breed, as they are in Canada, where they were first imported in 1881 and are known as Holstein-Friesians.

As the Friesian has spread around the world, some variation in type has arisen. At one extreme the Dutch Friesian is well-muscled, with short legs and a deep, wide body, so that it is useful for meat production in addition to milk. On the other hand the Holstein of North America has a much more dairy-like character and the emphasis is on high yields of milk and butterfat. In some countries there are separate Breed Societies for red-and-white Friesian cattle which occur from time to time in black-and-white herds.

The British Friesian is of intermediate type. The body is deep, and the hindquarters are long and wide, but the legs are longer than those of the Dutch Friesian. The udder is capacious. Mature cows weigh about 600 kg, and yield on average 4825 kg of milk with a butterfat content of 3·68%. At present the Friesian and Holstein are pre-eminent throughout the world for milk production. The colour is usually black and white in distinct patches. The head is long, and the short horns curl forwards and inwards.

Galloway bull

Galloway

During the 17th and 18th centuries as many as 30,000 cattle were sent each year down the drove roads from Scotland to the fattening pastures of England, mainly in East Anglia. Most of these cattle were Galloways, an old breed that belonged to the type of black cattle that inhabited the western regions of the British Isles. There is also evidence in the Galloway breed of the influence of Scandinavian cattle which were introduced by the Vikings. The Galloway Cattle Society was founded in 1877 and a separate Herd Book was published. Previously the Galloway had shared the Polled Herd Book with the Aberdeen Angus. The Belted Galloway is a separate variety which has its own Breed Society, but which is identical to the Galloway, except that a white-coloured belt completely surrounds the body between the shoulders and the hips.

The most common coat colour is black, but there is also an appreciable number of dun animals. Under natural conditions Galloway cattle grow a long coat of hair, but they also have a dense mossy undercoat which provides excellent protection against a harsh climate. The breed is polled and the head is short and broad. The body is deep and compact with well-sprung ribs. Mature cows weigh on average 480 kg.

The Galloway is a specialist beef breed which, because of its hardiness, is adapted to live in upland and exposed areas of poor fertility. It is able to utilize rough hill grazings and can be wintered out of doors on land where most other breeds would not survive. In northern England and south-west Scotland the Galloway is used widely to cross with Whitebred Shorthorn bulls to produce the Blue-Grey, which is one of the most popular and efficient crossbred suckler cows in Great Britain.

Gloucester

Gloucester cattle have been a minority breed since

Gloucester cow

they were overwhelmed by the explosive success of the Shorthorn in the 19th century. Several times they have been rescued on the brink of extinction, most recently by the formation of a Breed Society in 1973. The breed numbers about 60 cows and only three distinct bloodlines can be found among the bulls.

The Gloucester is a dark brown colour with a white stripe running down the back, hindquarters and tail, and along the belly. Bulls have a darker head and legs, and a well-developed crest. The horns are fairly short.

This breed has not been found to any significant degree beyond the limits of its native county. Previously milk from Gloucester cattle, because of the small size of the fat globules, was used in the manufacture of Double Gloucester cheese.

Guernsey

Cattle imported to the Channel Islands from Brittany by Breton monks in AD 960 formed the basis of the Guernsey, and remnants of the parent stock are represented still by the Froment de Léon cattle in France. In the 11th century a larger type of brindled cattle was introduced and crossed with the cattle imported earlier, to evolve the modern Guernsey breed. In 1789 and 1802 laws were passed to prevent further importations of cattle into Guernsey, so that the stock remained pure. Cattle were exported to England in the 18th century and the English Guernsey Cattle Society was founded in 1884.

The usual colour of Guernsey cattle is red and white, but in recent years the amount of white has decreased. The red ranges from fawn to dark red, and occasionally brindle or black-and-white animals are found. The head is long and the horns short and curved. The body is lean with definite dairy characteristics and a well-developed udder. Mature Guernsey cows weigh about 455 kg and yield on average 3660 kg of milk per lactation with a butterfat content of 4.5%.

Guernsey cows are noted for their longevity, and lifetime production of milk is an important measurement of merit within the breed. Calves weigh about 32 kg at birth, but the Guernsey is a specialist dairy breed, so that growth rates and carcase quality are relatively poor. The carcase fat is yellow, which does not attract the consumer, but on the other hand the rich colour of the cream is an advantage, while the large fat globules in the milk make it very suitable for butter making.

Until the early years of the 20th century a similar but separate breed existed on the neighbouring island of Alderney, but it has now been absorbed into the Herd Book for the cattle of Guernsey and has become extinct as a distinguishable breed.

Hereford

The quality of the cattle of Herefordshire was noted as early as the beginning of the 17th century, and 150 years later Marshall thought that they were the leading breed in Great Britain. They belong to the middle-horn group of breeds and are thus related to Devon and Sussex cattle. Cattle imported from Holland played a part in forming the breed and gave the Hereford its white head, a feature which it shares with the Blaarkop, a breed from Groningen in Holland. The first major phase of improvement of the Hereford breed occurred during the 18th century, under the guiding hand of the Tompkins family, who emphasized both the hardiness and early maturity of their

Guernsey cow

Hereford bull

cattle. At this stage the colour of the cattle varied a good deal from light grey, through grey and roan to red with a white face. During the 19th century the latter colour prevailed and was fixed by selection to become a hallmark of the breed.

The modern Hereford is a deep red colour with white markings extending forward from the withers to cover the head and brisket, the underparts of the body, the lower parts of the legs and the tassel of the tail. The head is short and broad, and an increasing number of Hereford cattle are now polled. The body is strong and deep and well covered with meat. The legs are short with strong bones. Bulls at 400 days of age weigh 440 kg and have a withers height of 111·5 cm.

The Hereford Herd Book was first published in 1846 as a private venture, but in 1876 the Hereford Herd Book Society was formed. The breed has now spread so widely that it has become one of the dominant breeds of the world and has adapted itself to a wide range of conditions from the icebound regions of Canada in winter to the sparse vegetation of the Australian bush or the South African veld.

Highland

The history of the Highland cattle of Scotland traces back at least to the 16th century when a definite breed of cattle was recorded in the Hebrides, but when the breed started to be improved in the 18th century two types emerged. On the islands a more primitive type existed, known as Kyloes. They were small, shaggy, and mainly black in colour, compared with those on the mainland which were very variable in colour, ranging from yellow through red and brindle to black. The Highland of the present day has

Highland cow

evolved from the fusion of these two types. The Highland Cattle Society was founded in 1884 and the first volume of the Herd Book was published the following year. Animals have been exported to all parts of the American continents, and to South Africa and Australia.

Highland cattle are picturesque animals with long shaggy hair over all the body, and falling in a thick fringe down the face. They have also a dense undercoat of soft hair which is exposed in summer when they shed their outer coat. The hair is most commonly red or brindle, but yellow and black animals are now becoming more popular. The head is short and broad, and the long horns on the cows sweep out and upwards to give them a flamboyant air. The body is broad and deep, and the legs are short, although a slimmer, long-legged beast is preferred for the herds which are kept under true mountain conditions. A mature cow weighs about 500 kg in normal breeding condition on the hill, and stands about 106 cm high at the withers.

The Highland is an extremely hardy, slow-maturing animal. Heifers calve first at $3\frac{1}{2}$ years of age, but are noted for their longevity, as are the bulls. They are crossed most commonly with Beef Shorthorn bulls, and in recent decades this cross has been fixed by selection to create a new breed, the Luing.

Jersey

Two waves of cattle migration met in the islands of the English Channel. A well-defined migration from Asia, which influenced breeds of cattle throughout southern Europe and across France, reached the Channel Islands and finally petered out in south-west England. At a later stage the Vikings spread the influence of Scandinavian cattle over a wide area, and the Jersey owes part of its ancestry to both these sources. The breed was valued highly in its native

Jersey cow

island, and in 1763 the importation of cattle was prohibited by law in order to preserve the purity of the breed. In 1833 the Royal Jersey Agricultural Society was formed to promote the breeding of the cattle, and in 1866 the Herd Book was established. Jersey cattle were exported to England as early as the end of the 18th century, and in 1878 the English Jersey Cattle Society was formed. The Jersey Cattle Society of the United Kingdom now deals with all registrations.

The Jersey has been exported widely and has proved particularly valuable for improving the native cattle in tropical regions, where its high heat tolerance is a valuable factor. It is an early maturing breed, and heifers usually calve at about two years of age. Cows yield on average 3458 kg of milk per lactation with a butterfat content of 4·9%. The fat globules are large, making the milk very suitable for butter production. Mature cows in England weigh about 390 kg, but are smaller on their native island. The bulls frequently have an unpredictable temperament.

The colour of Jersey cattle varies from light fawn to dark red and mulberry, with a light-coloured ring around the nose and on the underparts, while the hair on the head and neck is a darker shade, especially on bulls. White markings are uncommon. The head is small and fine, with a dished face that is deer-like in quality. The whole conformation is lean and typically dairy in type. The legs are short and fine-boned, and the udder is large and well developed.

Kerry

Although now restricted mainly to the south-west of Ireland, the Kerry was the dominant breed in that

Kerry bull

country little more than a century ago. It is the most typical and direct descendant of the native cattle of the western seaboard of Europe that were pushed to the outer fringes of the British Isles by later invasions and tribal migrations. It is thus related to the Kyloe, Welsh Black and the cattle of the Camargue. It has been introduced into the United Kingdom in small numbers, but has never established a firm foothold there. Kerry cattle were first shown at the Royal Dublin Show in 1844, and the Kerry Herd Book Society was formed in 1882, publishing the first volume of the Herd Book in 1890.

Kerry cattle are black, sometimes with a little white on the udder. In the mid-19th century various colours were found including brown and extensive white markings, but persistent selection established black as the dominant colour, except in small groups such as the line-backed Drimmon cattle. The head of the Kerry is light, with slender, graceful horns, which curve upward to give the cows an alert and attractive appearance. The conformation is lean and typically dairy in type.

The Kerry is adapted to tolerate cold and hard conditions in areas of low fertility, although it is also kept in the more fertile valleys. It is noted for its longevity, and for its high production in relation to its bodyweight. Mature bulls weigh up to 550 kg and cows about 370 kg. They produce on average 3250 kg of milk per lactation with a butterfat content of almost 4·0%. They are not of value for beef production when bred pure, and traditionally have been crossed with Beef Shorthorn or Aberdeen Angus bulls, with Charolais bulls being used more recently.

Lincoln Red

The original cattle of Lincolnshire were closely related to the red Angeln cattle of northern Germany, and were noted for their great size. They were valued highly in the late 18th century, but following Charles Colling's sale at Ketton in 1810 considerable numbers of Shorthorn or Teeswater cattle were taken to Lincolnshire and, crossed with the local cattle, formed the basis of the Lincoln Red Shorthorn breed. Although the Lincoln Red Shorthorn was included in Coates Herd Book when it was opened in 1822, the Lincoln breeders formed their own association in 1884 and published their first Herd Book in 1896. In 1925 the breed was amalgamated with the Shorthorn Society of Great Britain and Ireland, but established a separate identity yet again in 1941, and in 1960 the name was shortened to Lincoln Red.

The colour is a uniform, deep red. The head is

short and broad, and the short horns curl forwards and down. A polled strain has now been developed by using Red Angus bulls. The body is long, deep and well muscled, and the legs are short and strongly boned. The cattle are fairly large and at 400 days of age bulls stand 123 cm high at the withers and weigh 500 kg.

Originally the Lincoln Red was classed as a dual-purpose breed, but more recently greater emphasis has been placed on its beef production characteristics. Whereas in 1960 the average milk yield was 3600 kg per lactation and 3·7% butterfat, this has now fallen to 3425 kg per lactation with a butterfat content of 3·43%. However, this is still a very good level of production for a beef animal, and Lincoln Red cows rear their calves well. They are now used extensively in cross-breeding programmes for beef production.

Longhorn

The origin of the Longhorn is uncertain. It may be related to the Criollo cattle of Spain or it may be an indigenous descendant of the wild auroch in northern Europe. In any event it was first improved by breeders in western Yorkshire so that it was known as the Craven Longhorn. From there it spread to Westmorland and Lancashire and then to the Midlands. The original Longhorn was a good triple-purpose animal, but it was improved by Robert Bakewell who sacrificed many of its qualities, including milk yield, in order to produce a fatter, early-maturing beef animal. By 1800 the Longhorn was probably the most popular breed in the British Isles, but its popularity was short-lived and it had to give way before the relentless advance of the Shorthorn.

Lincoln Red bull

Longhorn cow

The breed derived its name from the distinctive long horns which sweep slightly backwards and outwards from the forehead, and then downwards, forwards or upwards in a variety of eye-catching styles. The horns are fine in texture, and should be free from any black colour. The hair colour varies from light red/roan through red to a dark plum-brindle, but all animals are line-backed—that is have a white stripe along the spine and down the tail. White hairs are interspersed on the head and face, and most animals have a white patch on each thigh.

The body is long and the legs are short. The Longhorn is popular with butchers because of the leanness of its carcase. At 400 days of age bulls weigh about 495 kg with a withers height of slightly more than 120 cm. Although milk production has been neglected, the butterfat content is extremely high and the milk was used at one time for cheese production because of the small size of the fat globules. The Longhorn is hardy and thrifty, and is a very docile animal.

Meuse–Rhine–Ijssel

The MRI, as it is usually called, evolved as a distinct breed in the east and south-east regions of the Netherlands, in an area bordered by the rivers from which it takes its name. Although it bears some resemblance to the Friesian it probably owes a good deal of its ancestry to the old indigenous cattle of the Netherlands, and is closely related to the red-and-white cattle which are found in the adjacent areas of Germany and Belgium. In 1906 the Netherlands Herd Book Society first registered them as a separate breed, but by 1970 the breed was attracting great interest from foreign breeders and was exported first to Great

Meuse-Rhine-Ijssel bull

Britain and then to North America. In Britain it was used by some breeders to increase the productivity of their Shorthorn cattle, and it has been used by breeders of Maine–Anjou cattle in France for the same purpose.

The performance standards of the MRI compare closely with those of the Friesian for milk production, but the MRI is superior for meat production. The body is wide, deep and heavily muscled, especially on the hindquarters. The rib cage is large, and the legs are short. The head is of medium length, and broad, and the short horns curve forward and inwards. The coat colour is red and white in distinct patches, but the legs, underparts and tassel of the tail are white, and many cattle have a white blaze down the face.

MRI cattle are robust and dual-purpose in type. They are thrifty and are able to adapt themselves to differing conditions and varying types of soil more readily than the Friesian. They produce a carcase of high quality with little undesirable fat below the skin.

Red Poll

Of the breeds of British cattle which have become extinct, the passing of the Suffolk Dun was the most regrettable. It was a dairy breed of high repute at the end of the 18th century, but it was not located near a centre of population, which would have ensured its continued existence. However, the Suffolk Dun was crossed with the red cattle in the neighbouring county of Norfolk, and from the mixture the Red Poll breed was developed. The hardiness and beef qualities of the Norfolk combined with the milking ability of the Suffolk gave the new breed a good dual-purpose appeal. The Herd Book was first published in 1873. The breed at that time was called the Norfolk and Suffolk Red Polled, but this has been shortened progressively to Red Poll.

The hair colour is red, and although the shade varies considerably, a deep red is preferred. The switch of the tail is white and some white may be found on the udder. The head is of medium length, broad and hornless. The body is long and deep, with well-muscled hindquarters. The legs are short. Mature bulls weigh about 725 kg and cows 520 kg.

Red Poll cattle have been exported to areas such as South Africa and Australia, but they have declined in popularity in Great Britain. The cows yield on average 3880 kg of milk per lactation with a butterfat content of 3·73%. In recent years Red Danish bulls have been used to raise the productivity of the Red Poll.

Santa Gertrudis

During the 20th century many attempts have been made in the USA to combine the high production

(*above*) Red Poll bull
(*below*) Santa Gertrudis bull

and quality of European breeds with the hardiness and tolerance of hot climates which Zebu (Brahman) cattle show. In several cases these crossing trials have resulted in the creation of a new breed, such as the Santa Gertrudis (Shorthorn and Brahman), Beefmaster (Shorthorn, Hereford and Brahman), Brangus (Angus and Brahman), and Charbray (Charolais and Brahman). The Santa Gertrudis was developed on the King Ranch in Texas, where the purebred European cattle, which replaced the original Texas Longhorns, were unable to tolerate the drought and parasites in that area. Thus in 1910 they were crossed with imported Brahman cattle and very early in the programme this policy produced an outstanding bull calf called 'Monkey', which became the foundation sire of the Santa Gertrudis breed.

Santa Gertrudis cattle today carry approximately five-eighths Shorthorn blood and three-eighths Brahman blood, with a suggestion of some Afrikaner influence. A deep cherry red has become the accepted colour, with no white markings. The head is of medium length and, in the bulls particularly, the horns sweep back from the head so that the forehead is distinctly convex. The horns of the cows grow outwards and are fairly straight. The Santa Gertrudis has inherited from the Brahman an abundance of loose skin and a small hump over the shoulders and neck. Otherwise the conformation is typically European with a long, well-muscled body carried on legs of medium length.

The Santa Gertrudis is valued mainly for its use in hot, dry areas. It is popular in the southern States of the USA, and has been exported to Central and South America, South Africa and Australia. It has proved most successful in Queensland. The Santa Gertrudis Breeders International was formed in 1951.

Shetland

Shetland bull

The native cattle of the Shetland Islands have existed there since records were made and thus their origin is a matter for speculation. It is likely that they belong to the Scandinavian group of breeds and they are similar in type to the Jersey. At various times in the past bulls of other breeds have been introduced, in particular Shorthorn and Angus, and have threatened to extinguish the Shetland by crossbreeding. The influence of Highland cattle has been noted also, while in 1923 two Friesian bulls were introduced. However, exposure of the cattle to adverse climatic conditions and low levels of nutrition ensured that only the hardy and thrifty native cattle could survive so that the influence of imported animals was insignificant. A Herd Book Society was formed in 1910, but the breed has declined steadily in numbers.

In the past a variety of colours could be found including black, dun, or black and white in various patterns. Since the introduction of Friesian bulls in

1923 black and white has been favoured and is now the only colour. During the early years of the present century mature Shetland cows weighed about 205 kg, but they have now increased in size, as a result of improved husbandry techniques, and weigh about 330 kg. They are short-legged but the barrel is large. The horns are tightly curved and slender, while the soft skin again makes them similar to the Jersey.

Shetland cattle have a low feed requirement and can tolerate low quality feed. They are dual-purpose animals which produce high levels of meat and milk in relation to their size. The udder is large and capacious, and they are able to fatten readily even on indifferent pasture. Their beef is noted for its tenderness and flavour.

Shorthorn

As early as the mid-16th century there existed in the north-east of England a race of superior short-horned

Dairy Shorthorn cow

cattle. They were mainly of Dutch descent, but they were influenced slightly by Scandinavian blood. Two centuries later they became prominent as a result of the expertise of breeders living in the valley of the Tees, so that the large-framed, high-yielding cattle which they bred became known as the Teeswater, but were also referred to as Durham, Holderness or Yorkshire. Although there were many eminent breeders of Teeswater cattle in the early 18th century, the visit of the Colling brothers to Dishley Grange, Bakewell's home in Leicestershire, in 1783 marked a significant point in the fortunes of the breed. As a result of this visit the Colling brothers adopted Bakewell's method of inbreeding (see p. 58) but carried it to even greater extremes.

The Shorthorn, as it was now known, spread throughout the whole of the British Isles and its influence was felt in many countries abroad, especially among the Charolais and Maine-Anjou cattle in France. Very early in this stage of development different varieties of the Shorthorn type emerged, and now there are four clearly defined breeds namely the Dairy Shorthorn, Northern Dairy Shorthorn, Beef Shorthorn and Whitebred Shorthorn, in addition to the closely related Lincoln Red.

Dairy Shorthorn

In 1822 George Coates founded the Coates Herd Book, in which all Shorthorn cattle were registered, but in 1905 the Dairy Shorthorn Association was established to promote those families which were bred for milk production. The Dairy Shorthorn is essentially a dual-purpose breed producing good beef steers, while the cows yield 4250 kg of milk per lactation with a butterfat content of 3·6%.

The colour is red, white, red and white, or roan. The head is of medium length with short horns which usually curl forwards and down. Recently the Shorthorn Society of Great Britain and Ireland has initiated a crossing experiment to stem the falling popularity of the breed by introducing bulls of other breeds such as the Red-and-White Canadian Holstein, Meuse-Rhine-Ijssel and Simmental.

Northern Dairy Shorthorn

In the relatively undisturbed dales of the Pennines in Yorkshire and Durham a distinct type of cattle was established known as the Northern Dairy Shorthorn. Evolved in the harsh surroundings of its upland habitat, it is a thrifty breed capable of producing good yields of meat and milk under difficult conditions. Cattle of this type were not registered, but were kept pure by the interchange of bulls between neighbouring dales, and they are a direct offshoot of the old Teeswater cattle, refined by a slight introduction of Ayrshire blood during the first half of the 20th century. The Northern Dairy Shorthorn Breeders Society was formed in 1944.

This attractive member of the Shorthorn family is a stylish animal with an alert carriage, horns curved and tilted upwards, and a classic dairy conformation. The most popular colour is light roan, but red, white, red and white, and other shades of roan are found. The average yield is 4180 kg of milk per lactation with a butterfat content of 3·56%.

Beef Shorthorn

Beef Shorthorn cattle were registered in Coates Herd Book, and it was not until 1959 that the Beef Short-

Northern Dairy Shorthorn cow

Beef Shorthorn bull

horn Cattle Society was formed, although its fame as a beef animal was established long before that time. It has been exported widely, being extremely popular in Argentina and Australia, and with a strong following in North America and South Africa. As a result of the demand for export, the Beef Shorthorn became extremely short-legged and compact with a resultant loss of growth rate. Bulls at 400 days of age weigh 410 kg, but are now being selected for greater size.

The colour range is the same as for the Dairy Shorthorn but dark roan is preferred. The body is broad and deep and extremely well muscled. The head is short and broad, and the horns are short and frequently curl downwards. Some strains are polled.

Whitebred Shorthorn

The Whitebred Shorthorn is a faster growing variety of Beef Shorthorn that is localized mainly in the Border Country of England and Scotland, and is used specifically for crossing with Galloway cows to produce Blue-Grey suckler cows. It is always white in colour.

Simmental

Red-and-white cattle have existed since the Middle Ages in the Bernese Oberland of Switzerland, and particularly in the Simme valley which gave its name to the Simmental breed. Until the middle of the 19th century little attention had been paid to the improvement of the cattle, and a variety of colours could be found including red, or black spotted, or whole red. In 1862 a Herd Book was established and the characteristics of the breed were identified. The

Simmental bull

Simmental adopted the red-and-white colour, but the black spotted pattern was retained in the Fribourg breed, which is a large, high-performance, dual-purpose breed found in western Switzerland, but which is now regrettably almost extinct.

Although they were very good work animals in the past, Simmental cattle are now used for the production of meat and milk, and are more widely distributed throughout Europe than any other breed. Each country tends to adopt its own name for these cattle so that the German Fleckvieh and the French Pie Rouge de l'Est are basically Simmental. They have also been exported in large numbers to North America.

The interest of foreign buyers has been directed mainly to the beef qualities of the Simmental, but it is also an excellent milk-producing breed. The average yield is 4140 kg milk with a butterfat content of 4·05%. Mature cows have a withers height of 136 cm,

South Devon bull

and weigh 730 kg, while mature bulls stand 143 cm at the withers and weigh 1090 kg. Bulls at 400 days of age weigh 545 kg and are heavily muscled, especially over the rump and thighs. The body is long and deep, and the legs are fairly short. The dewlap is well developed and there is much loose skin at the throat. The head is of medium length and the horns curve outwards and forwards, turning upwards at the tips. The hair colour is red and white in patches, but the red ranges from yellowish-brown to deep red. The head, legs and tassel of the tail are always white.

South Devon

The rich lowland plains in a small area in the south of Devon and Cornwall are the home of Britain's largest native cattle, the South Devon or 'South Hams'. Their origin is not known with any certainty, but there is little doubt that they were influenced to some degree by cattle migrations from sub-tropical regions. The South Devon Herd Book Society was formed in 1891, and the breed has been in demand for export especially to South America and South Africa.

The South Devon is a true dual-purpose breed. Although the size of the cattle has been reduced during the present century by selecting a shorter-legged type, they are still of massive proportions. Bulls at 400 days of age weigh 530 kg, although individual specimens may weigh up to 770 kg. They stand 126 cm high at the withers. The cows yield on average 3390 kg of milk per lactation, although individual yields in excess of 10,000 kg have been recorded. The average butterfat content is 4·14%, and this has contributed to the fame of Devonshire clotted cream.

South Devon cattle are long-lived and docile, although a few bulls can become vicious. The body is long and deep, but occasionally is rather narrow. The hindquarters are well muscled. The head is long and fairly broad, and the horns are short, frequently curving forward and downward. Some polled strains have been developed. The hair is of medium length, with a tendency to curl, and sandy-red in colour. There are no white markings.

Sussex

Sussex cattle were at one time a purely draught breed and were frequently seen in the yoke up to the end of the 19th century. They belong to the group of old red British breeds, being closely related to the Devon, and are derived from the red, medium-horned cattle which are known to have existed in the south-east of England for at least 300 years. The Sussex Cattle Society was formed in 1874 and the first volume of the

Herd Book was published in 1879. Sussex cattle have been exported to many countries and have proved particularly popular in South Africa, for crossing with the Afrikaner, and in the southern states of the USA. They are not susceptible to the eye irritation that affects most other British breeds in hotter and drier climates.

The colour of the cattle is a deep cherry red, with only the switch of the tail being white. The head is short and broad, and the medium-sized horns spread forwards and upwards. Since 1966 polled Sussex cattle, developed by several leading breeders in a controlled programme, have been eligible for entry in the Herd Book. The body is long and deep, but until recently the relatively heavy forequarters served as a reminder of the use of these cattle as plough oxen.

The Sussex is a meat-producing animal, but a good proportion of cows yield too much milk for one calf. The breed does not quite compare with the Devon for growth rate, as bulls at 400 days of age weigh 445 kg and have a withers height of 117 cm, but it does possess other valuable characteristics. The carcase is relatively lean and of high quality, while the adult cattle are excellent grazers and are able to convert efficiently a wide range of herbage types from the lush pastures of the Pevensey Marsh to the sparse vegetation of the veld of South Africa.

Welsh Black

As the ancient Britons retreated to the western fringes of the British Isles before the invading Saxons they took with them their cattle, which were the ancestors of the modern Welsh Black. By the beginning of the 19th century two main types existed, the North Wales or Anglesey type, and the Castlemartin

Sussex bull

which was found mainly in the south-west of Wales. In 1904 the two types were amalgamated to form the Welsh Black Cattle Society. The Anglesey emphasized meat production while the Castlemartin was more of a dairy type, so that the modern breed is dual-purpose in character, although some variation in type still exists.

The Welsh Black has been considered by most authorities to be a hill breed, and it does show considerable hardiness and thriftiness under upland conditions, but recently it has shown that it can achieve high levels of production when given suitable opportunity. Meat production now receives the greatest emphasis and bulls at 400 days of age weigh 475 kg and stand 119 cm high at the withers. The carcase contains a high percentage of lean meat. Cows yield on average 3100 kg milk per lactation, with a butter-fat content of 3·9%, and they are noted for the persistency of production and of longevity.

Welsh Black cow

The colour is black and, apart from a few white hairs on the udder, white markings are not allowed. One hundred years ago white colour patterns occurred frequently. Today examples exist of Welsh Black cattle which are either belted, or exhibit the White Park colour pattern, and are known as 'White Welsh'. Very occasionally red Welsh Black cattle are seen. The head is short and broad, and the horns are strong with black tips. These sweep outwards and forwards, and are most nearly comparable to those of Highland cattle. The conformation is adapted to living on upland pastures and is not of the same blocky compactness that is typical of the extreme beef breeds.

White Park

About 700 years ago much of Britain was covered in large areas of forest. At that time wild cattle were prized beasts of the chase, and when Henry III of England granted a Charter to certain of his nobles to enclose extensive tracts of forest as hunting parks, it was natural that the wild cattle should be enclosed along with other wild beasts. Some of these old herds still exist, and those at Chillingham, Dynevor and Hamilton remain in their original park. If their history is traced back further it reveals a relationship with the white breeds of Italy and the Grey Steppe cattle of the south-east of Europe.

A Breed Society was formed in 1918, but this included both White Park and polled British White cattle, and in 1946 White Park cattle ceased to be registered by this Society. No registration facilities existed until 1974, when Countrywide Livestock

White Park bull

opened a Herd Book, but detailed records were maintained by some herds during the interval. The breed is very low in numbers, although interest in these cattle and the demand for breeding stock is increasing rapidly.

The White Park is a distinctive and attractive animal. It is white in colour, with black ears, nose, eyelids, feet and teats. The head is of medium length and broad, with long horns which sweep outwards, forwards and upwards. The skin is fine and may be slightly pigmented, as befits a breed which originated in a hotter climate, but it has adapted well to British conditions.

Possibly because most animals are inbred to a relatively high degree, the performance of purebred White Park cattle is only average. Bulls at 400 days of age weigh 460 to 470 kg. But their crossbred progeny show considerable hybrid vigour, thus making the White Park a valuable beef sire. The cows have strongly developed maternal instincts. They are able to thrive on low quality feed.

HORSES

Belgian

Heavy horses were being bred in Belgium in the 11th century and the heavy Belgian horses of the present day are considered by many authorities to be a direct descendant from the massive horses of Quaternary Europe. Without doubt the Belgian Heavy Draught, or Brabançon as it is sometimes known, can trace its history back to the Middle Ages when horses of this

Belgian Heavy Draught Horse

type were exported to many other European countries where they exerted a great influence on the breeding of heavy horses. Later, around 1850, they were exported to North America and have become popular in the Mid-West. Belgian horses were imported into Britain in considerable numbers immediately before the Second World War, and the breed is related to the old Flemish horse which contributed to the formation of the Shire and Clydesdale breeds. A Stud Book was commenced in 1886.

The colour is normally sorrel, dun, or roan, but horses in the USA frequently have a white mane and tail, white feet, and a white blaze on the face. The American horses are also larger than those in Belgium. The mares stand 16–17 hands and weigh 770 kg, while the stallions are a hand higher and weigh about 135 kg more.

Despite its great size the Belgian Heavy Draught is active and moves with a good action. It is an exceptionally strong and willing worker, and has a docile temperament. The head is generally small and well shaped, but the shoulders are massive and the strong, compact body is carried on short, well-muscled legs, which carry 'feather' on the fetlocks.

Cleveland Bay

Yorkshire has long been a noted centre of horse breeding and the Cleveland Bay is a product of this county. Prior to the era of coaches, travelling salesmen were called chapmen and the horses that they used to carry their merchandise became known as Chapman horses. These were the foundation of the Cleveland Bay, which were improved in the mid-18th century by the introduction of two horses of mainly Arab breeding. Before this time, in the reign

Cleveland Bay Horse

of Queen Elizabeth I, they filled the demand for strong coach horses. The Yorkshire Coach Horse was developed as a specialized offshoot of the Cleveland Bay for this purpose, but the two types were reunited in one Stud Book towards the end of the 19th century. Today the Cleveland Bay is used for many purposes. It is in great demand either to draw the royal coaches in London or for use in driving competitions, which are becoming increasingly popular. Its natural ability to jump has brought Olympic medals for show jumping.

The Cleveland Bay stands about 16 hands high. The head is large and carried proudly on a long neck. The body is strong, but rather long, and is carried on hard, short, clean legs, with 22·5 cm bone below the knee. The colour is bay or bay/brown with black points. A small white star is permissible.

The Cleveland Bay was exported to North America in 1820 and the Cleveland Bay Society of America was formed in 1885. It is crossed with the Thoroughbred to produce hunters, and with range mares to produce reliable cow horses. It was used also in the formation of the Clydesdale and Oldenberg breeds. It has an excellent temperament, and the longevity, and hard-wearing qualities of the breed are recognized in many parts of the world.

Clydesdale

The Clydesdale was established as a breed during the first half of the 18th century, in Lanarkshire. It was based on the interbreeding of the native heavy horses, but English and Flemish stallions were used. The foundation sire of the breed was a black Flemish stallion which was introduced about 1715, but most pedigrees trace back as far as 'Glancer' who was foaled about 1810. At a later stage, further crosses with the Shire were made together with a limited introduction of Cleveland Bay blood. The numbers of the Clydesdale breed increased steadily during this period, and in 1877 the Clydesdale Horse Society was formed and published its first Stud Book.

The average height of Clydesdale stallions is 17·2 hands and of mares 16·2 hands. The conformation is less massive than the Shire. The body is not so deep and wide and the legs are longer. The 'feather' is confined to the back of the leg and the head is more refined. The colour is bay, brown or black, and only rarely grey or chestnut. White faces and legs are common, and white markings on the belly or white hairs mixed throughout the coat occur more frequently than before.

Among draught horses the Clydesdale is noted

Clydesdale Horse

particularly for its freedom from leg troubles, which gives the breed a long working life. It excels more for speed, agility and wearing qualities than it does for weight and strength. It has a more nervous temperament than the Shire, and its narrowness of build sometimes results in an inability to maintain its physical condition when working under adverse conditions of nutrition and climate. On the other hand it possesses a harder, flatter bone than the Shire, and it is claimed with justification that this breed possesses quality and weight without excessive bulk.

Dales Pony

During the 18th and 19th centuries lead mining was an important industry in the Pennine hills of North Yorkshire and County Durham, and pack-horses

Dales Pony

were used to carry the lead from the mines to the coast. The local ponies were used for this purpose, carrying loads of 200 kg and travelling up to 400 km per week. Developed in such a severe testing ground, the Dales Pony, like its near relative the Fell Pony, is strong, active and sure-footed. Originally both breeds shared a common ancestry, being descended from the 'Celtic' pony, but the Dales is now about one hand higher, probably due to the introduction of a limited amount of Clydesdale blood.

The Dales Pony is a native of the Pennine dales, and when lead mining ceased in that area, it became the main source of power on the small Dales farms. It has a good temperament, is very largely free from inherited weaknesses such as ringbone, spavin, etc., and is extremely hardy. It can be regarded as a small draught horse.

The breed standards allow a height of up to 14·2 hands, but many specimens exceed this size. The body is strong, full and well muscled. The feet and legs are sound with the heels covered in 'feather'. The head is stronger than in many pony breeds and the neck may be rather short. The most usual colour is black, but bay and brown are also common. Some animals have a white star or white feet.

Exmoor Pony

The Exmoor Pony is probably the most typical modern representative of the 'Celtic' pony that was bred by the inhabitants of Great Britain before the Roman invasion and used by them to pull war chariots in the uprisings and battles that followed. Since that time

Exmoor Pony

the breed has evolved on that wild tract of moorland in the western part of Somerset, from which it took its name. Attempts have been made to 'improve' the Exmoor by increasing its size. In the 19th century Dongola Arabs were used in crossing trials, but the experiment was a failure as the progeny lost type and hardiness. On the other hand the Exmoor has made significant contributions to the horse breeding industry in other fields. The Colonel, winner of the Grand National in 1868, was by a part Exmoor stallion.

The Exmoor Pony is generally bay or brown, darker on the back, but merging into a lighter mealy colour under the belly, and inside the forearms and thighs. There should be no white markings. The mealy nose is a distinctive point of the breed, as is the mealy coloration of the eyelids ('cingle') and the pronounced eyebrow known locally as 'toad eye'. Mares should not exceed 12·2 hands in height and stallions 12·3 hands. The Exmoor has a wide forehead, broad nostrils and short, thick ears. The legs are clear and short with good bone. Every registered animal carries the brand of the Breed Society's star and its herd number on the near shoulder, and its individual number on the near flank.

The great quality of the breed is its hardiness and thriftiness. With their history as pack ponies, they are bred to carry weight and they can maintain their condition on small quantities of feed. In fact, if too highly fed, they are apt to become too fat. Currently the Exmoor Pony has a wide range of uses. It makes an excellent children's hunter and cross-country driving pony. Its weight-carrying ability and independent spirit suit it well for pony trekking, and for generations farmers on Exmoor have used these ponies for shepherding.

Percheron

The most widely dispersed heavy breed of horse in the world is the Percheron, which derives its name from the old French district of La Perche. Its origin can be found in the working horses of northern France, but it was influenced to some extent by the introduction of Arab blood, and while this had the effect of refining the type, it caused variations within the breed which still exist. The Percheron has been exported widely and is very popular in North America. It was first introduced into Great Britain during the First World War, and the British Percheron Horse Society was formed in 1918.

The Percheron is one of the smallest of the heavy draught breeds. The height averages 16 hands for mares and 16·3 for stallions. It has a short, compact body of great depth, but the head is small, full of character, and refined for a heavy breed. The prominent eyes and slightly dished face betray the Arab

Percheron Horse

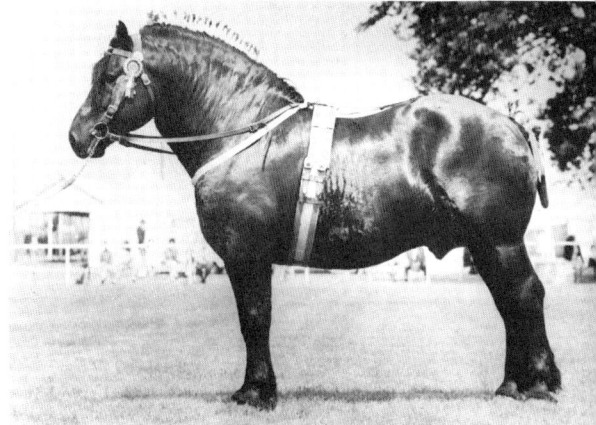

influence in its ancestry. Although it is short-legged and carries strong bone, it is a very active horse with great stamina in the trot. Grey and black are the only colours seen, and the legs are devoid of 'feather'.

For generations French draught horses have been worked on rough stone roads, and the soundness of the hard blue hoof of the Percheron resulted from working on this surface. The popularity of the Percheron is based on its clean legs, good temperament, soundness, and good stamina in fast work.

Quarter Horse

More than 300 years ago, during the colonial era in Virginia, horse-racing was carried on along country lanes near plantations or on village streets, where distance was limited. For this type of racing, horses with great acceleration were required, and as they rarely raced beyond a quarter of a mile the breed became known as the Quarter Horse. It was based on a mixture of the English Thoroughbred and stock of Spanish origin from Florida, but the first stallion of importance was 'Janus' who travelled in Virginia and North Carolina between 1756 and 1780. Although he was noted for his stamina when he raced in England, his progeny in America excelled in short distance races. 'Janus' stood 14·2 hands high, but it was the powerful muscling of his hindquarters that was transmitted to his progeny.

Very early in its history the Quarter Horse was adapted by ranchers and cowboys as a roundup and trail driving horse, for it possesses inherent 'cow-sense', and its acceleration and manoeuvrability make it an invaluable asset when handling cattle. In 1941 the American Quarter Horse Association was founded, and since that time the breed has spread

Quarter Horse

rapidly through North and Central America, and many countries overseas. The great popularity of rodeos in North America provides yet another purpose for the Quarter Horse, for it is claimed that it can do more jobs better than any other horse in the world.

The height of Quarter Horses varies from about 15·2 to 16·1 hands and the weight from 550 kg to 590 kg. The head is short and broad, with wide-set, kindly eyes that indicate its docility and excellent calm temperament. The back is closely-coupled, the shoulders are strong and sloping, and the hindquarters are broad, deep and powerful. The forearms and gaskins are particularly well muscled. The Quarter Horse can be seen in the full range of solid colours. White markings on the body are not allowed, but a white star or blaze and white feet are common.

Shire

Under a variety of names such as the Great Horse, Old Black English Horse and War Horse, the ancestors of the modern Shire have been bred for centuries in the heavy-soiled districts of Lincolnshire and Cambridgeshire. The Shire resulted from crossing stallions, imported from Flanders, with the native English mares. The Great Horse was valued for its ability to carry armoured knights in battle, and kings of England from the time of John have encouraged the breeding of larger and improved animals of this breed. It was not until the 18th century that the horse effectively superseded oxen as the source of power on farms. The Breed Society was formed in 1878, but more recently Clydesdale blood has been used, especially to breed a type with cleaner legs.

The modern Shire is probably the heaviest of all the draught breeds. In show condition stallions may weigh 1100 kg, but when in working condition 940 kg would be the average weight. The head is heavy and rather coarse, and the body is short, wide and very strong. The forearms and thighs are of great size and the bone should measure 27·5 cm below the knee. The legs carry a considerable amount of 'feather' below the knee and the hock, not only at the back of the leg, but also at the sides. Although this feature was most attractive in the show ring, it was very difficult to keep clean on working horses on arable farms.

Shire stallions on average stand just over 17 hands high and mares about 16·3 hands, although occasional specimens attain a height of 18 hands. The great advantage of the Shire is its strength and, although it is probably the slowest worker among the heavy breeds, it is steady, reliable and does not shirk its duties. The most common colours are bay, brown

Shire Horse

and black, with grey, chestnut and roan less prevalent. At one time piebald and skewbald specimens were found, and extensive white markings on the legs and face are usual.

Suffolk

In the 16th century writers referred to the Suffolk horse as the 'old breed', and it can claim with justification to be the oldest of the heavy breeds. The most important stallion, to which all Suffolk horses trace their descent in direct male line, was Crisp's Horse, who was foaled in 1760. Other stallions were drawn from a variety of sources with the Belgian heavy horse at one extreme, and trotting horses at the other extreme. As a result of this varied ancestry the Suffolk

Suffolk Horse

Punch possesses a unique combination of characteristics, and the breed has now spread far outside its native area, although it is not so widely popular as some of the other heavy breeds.

The colour is always chestnut, although this may be light or dark and vary from yellow chestnut to liver chestnut. Sometimes there is a white star on the forehead, but no other markings are seen. The average height is about 16 hands, but the body is powerful and muscular, and the legs short, so that the Suffolk Punch is heavier than its height might indicate. Stallions fitted for show weigh 1050 kg, and 900 kg when in working condition. The legs are almost free from 'feather' and the feet, which used to be a weak point, have been much improved.

Unlike most of the heavy breeds the Suffolk Punch

moves well at the trot. At the same time it is noted for its staunchness and courage in harness, and its pulling power was well recognized as a result of weight-pulling contests held in East Anglia. In these contests teams of horses were matched against each other or were forced to pull very heavy loads, and while the Suffolk committed its total energies to the task, many willing horses were ruined by this pastime. Another outstanding merit of the Suffolk Punch is its constitution. It can tolerate exposure and scarcity better than other breeds, and its ability to thrive under poor conditions was very evident during the First World War when these horses were used to draw gun carriages. When well fed they put on a great reserve of fat and can be kept easily.

PIGS

The Berkshire

During the latter part of the 18th and in the early part of the 19th centuries many pigs of Asian origin were imported into Britain, either directly from China and Siam, or indirectly through Mediterranean types, particularly the Neapolitan. The British breed which owes most of its ancestry to these imported pigs is the Berkshire, which originated in the valley of the Thames. It inherited the Chinese characteristics of early maturity, a short head with a dished face, and a short, wide, compact body.

Initially the Berkshire was reddish brown in colour with black patches, and 'large ears hanging down over their eyes', although descriptions vary widely. The modern breed is dark in colour, almost black, with white points, that is tail, snout and feet,

Berkshire sow

and it has prick ears.

The early maturity of the Berkshire makes it suitable for the pork trade, and at one time it was supreme in Show classes for pork carcases. It has been exported to many areas including Australasia and North America, and in the USA contributed to the formation of the Poland China breed. It is at present low in numbers, but is attracting interest from South-East Asia and the Pacific regions.

British Saddleback

The British Saddleback is a recent development which resulted from the amalgamation of the Essex and Wessex breeds. The Essex originated in East Anglia and the Wessex in Dorset. Pigs of this type were described in southern England in the first part of the 19th century, and probably formed the population from which the Hampshire was imported into North America at about that time.

All three breeds exhibit the same colour pattern. They are black with a white belt or saddle over the

British Saddleback sow

shoulders. In addition, the Essex has white hind feet and a white tail. Pigs with this colour pattern do not breed entirely true to type, and the colour of the progeny may range from all-black to a wide white belt covering almost the whole body. The British Saddleback has lopped ears, in contrast to the prick ears of the Hampshire.

The British Saddleback is a dual-purpose breed. It is hardy and suited to outdoor systems. The sows milk well and rear good strong litters.

Chester White

The Chester White breed originated in Chester county in Pennsylvania, and thereby obtained its breed name. It was derived from crosses between three white breeds from England, namely the York-

Chester White gilt (young female)

shire, Lincolnshire and Cumberland. The Yorkshire, or Large White, is now a breed of international repute, but both the Lincolnshire Curly Coat and the Cumberland are extinct. The process of evolving the Chester White began about 1815. In 1884 the first breed association was founded, but as these pigs became popular and spread across North America, several other recording associations were started, and it was not until 1911 that all these organizations were combined into the Chester White Swine Record Association.

The Chester White is a large pig. Before 1900 mature pigs weighed slightly less than 300 kg, but during the present century the trend towards a larger type has resulted in some boars weighing in excess of 450 kg. Pigs of this breed are always white, but they have blue spots, or freckles, in the skin. They have semi-lop ears and a snout of medium length. The modern Chester White is noted for its ability to produce a high quality, lean carcase, but at the same time the sows are excellent mothers and produce large litters of pigs.

Gloucester Old Spot

Known previously as the 'Orchard Pig', the Gloucester Old Spot is an old local type that has been established for many years in the Berkeley Vale, but which was little known elsewhere until the Breed Society was established in 1914. It has never spread to any extent beyond its native area, and is still one of the numerically small breeds in Britain.

The Gloucester Old Spot is a useful grazing breed, and it is able to live on windfalls in the many orchards which are found in the Severn valley. It is hardy and the sows are good mothers. Originally pigs of this

Gloucester Old Spot sow

breed were white with a liberal covering of black spots, but in recent years breeders have selected those with fewer black spots so that many pigs today are almost entirely white.

Landrace

Landrace means native breed, and refers to a type of pig that is found around the north-western edge of Europe. It includes the pigs of Sweden, Norway and Denmark, the Welsh, and the British Lop, known previously as the Long White Lop-eared, from south-western England. These breeds are closely related and similar in type, being white, long-bodied and lop-eared.

The Scandinavian breeds have been selected for many years for bacon production, and this process has been carried to such an extreme that although the pigs are long and lean, they suffer from weak legs and sometimes produce meat of lower quality. The

Danish Landrace hog (castrated male)

selection of the two British breeds has been intense, and they have retained to a greater degree the qualities of hardiness and thriftiness. The British Lop has remained a localized breed, but the Welsh has spread throughout the British Isles, while the Landrace breeds from Scandinavia have acquired an international reputation.

Large Black

The foundation stock for the Large Black breed was drawn mainly from East Anglia and south-west England, but pigs of this type had been in existence for many years before the formation of the Breed Society in 1899. Large Black pigs have been exported widely, especially to warmer countries.

They are large pigs and mature boars weigh up to 500 kg when fat. They have a relatively heavy coat of black hair, which is fine in texture, but the skin is not deeply pigmented. The lopped ears are

Large Black boar

long, and the tips reach the nose, so that pigs of this breed do not have a clear vision, which may partly account for their reputation for docility.

The Large Black is a dual-purpose breed being intermediate in type between the Large White and the Berkshire. The sows are noted for their hardiness and for their ability to thrive on low levels of feeding. Although they are not quite so prolific as the Large White, they are excellent mothers and efficient converters of low quality feed.

Large White

In the first half of the 19th century factory workers in the West Riding of Yorkshire evolved a breed of pig that was based on the early North Country type and which contained very little Chinese or Neapolitan blood in its ancestry. It first attracted widespread interest in 1851 when Joseph Tuley, a weaver from Keighley, exhibited an improved animal at the

Windsor Royal Show. It is now the most popular breed in Great Britain and has been exported to many other areas including Australasia and North America, where it is known as the Yorkshire.

It is a pig of great size. Mature boars weigh 355 kg in breeding condition and up to 510 kg when fully fat. The breed standard requires a white skin and coat, but some animals show small areas of darker skin. The head is of medium length and the face is slightly 'dished'. The ears are pricked. The body is long and smooth.

Large White pigs are used mainly to produce bacon. They are late maturing and possess a lean carcase. The sows may be rather temperamental on occasion, but they are prolific breeders and have good milk production.

Middle White

The Middle White breed was evolved by crossing Large White and Small White pigs. The Large White is described elsewhere in this book, but the

Large White sow

Middle White sow

Small White, which is now extinct, was a small, excessively fat animal, with fine bones and a short face, that owed most of its ancestry to imported Chinese pigs. The Middle White was recognized officially as a breed in 1882, although the type was not firmly established at that time. It enjoyed a period of popularity during the first half of the 20th century, but is now rare and in danger of extinction.

It is a relatively small breed with mature boars weighing 285 kg in breeding condition and 405 kg when fully fat. It has prick ears, and its head is short and excessively dished. The conformation of Middle White pigs is short, wide and compact. They have short legs and heavy jowls, and are white in colour.

They are early maturing pigs and are very suitable for the pork trade. The sows are reasonable mothers, but Middle White pigs are used most often in crossbreeding programmes.

Poland China

The Poland China breed was developed during the first half of the 19th century in south-western Ohio, or more specifically in Warren county. It was derived from a wide mixture of types of swine, including 'China' pigs from Philadelphia and old Berkshire pigs, which were crossed with local types. After about 1846 no further pigs were imported and the new breed became known as the 'Warren County Hog'. It was not until 1872 that the name of Poland China was adopted officially. A recording society was established in 1876, and since that time more than two-and-a-half million head have been recorded, to make the Poland China the most numerous breed in the USA during the 20th century.

Pigs of this breed are black in colour, but normally have six white points, that is legs, nose and tail. They have semi-lop ears and a snout of medium length. Poland China sows grow to a great size and

Poland China boar

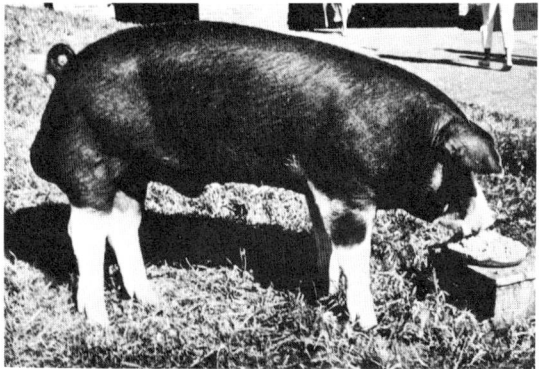

are highly productive, but the main value of the breed is the ability of the pigs to grow quickly and produce a high quality carcase with a high percentage of lean meat.

Tamworth

The Tamworth evolved from a mixture of imported Red Barbadan or Axford pigs with the old British pig, which was an indigenous descendant of the European wild boar. The Tamworth derived its attractive red-gold colour from the Axford pig, but many other characteristics, including the long snout are reminiscent of the wild boar. However, the sows are docile, being good mothers and not clumsy with their litters.

The lean quality of the carcase is a quality for which the breed has always been noted especially in North America, where it was valued particularly for bacon production, while in England it filled a specialist role in the Midlands as a pork pig.

Tamworth sow

The Tamworth has been exported widely, especially to countries with a hotter climate, as the pigs do not sunburn like the white breeds. Thus they are more important in Australia and North America than they are in Great Britain, and they are at present proving very suitable for use in South-East Asia. On the other hand they are equally able to thrive in outdoor systems in Britain, being eminently suitable for reclaiming scrub and rough pasture.

GOATS

Nubian

The Nubian goat originated in Nubia in north-eastern Africa, and it is reputed that in 1860 Nubian goats which the King of Abyssinia presented to Napoleon III were sent from Africa to Europe to supply milk for a young hippopotamus. Purebred animals of this breed did not thrive in England and they were crossed with the native short-haired British goats. The crossbred progeny became a new breed called the Anglo-Nubian and were exported to North America about 1900. They are now the most popular breed of milk goat in the USA.

In contrast to the other breeds of milk goat, the Nubian has a strong Roman nose and prominent forehead. The ears are long and pendulous, and the hair is very short. A wide variety of colours are allowed in the Nubian, but black, red or tan are the most common colours, any of which may be combined with white. Adult females stand over 70 cm at the withers and weigh 55 kg, and adult males weigh 75 kg. The Nubian produces less milk than the Swiss breeds, but the milk is very rich, and the breed is often referred to as the 'Jersey of milk goats'.

Schwarzhal

The Schwarzhal goat originated in the Swiss Canton of Valais and is found also in the Rhône Valley. During the 14th century returning Crusaders brought

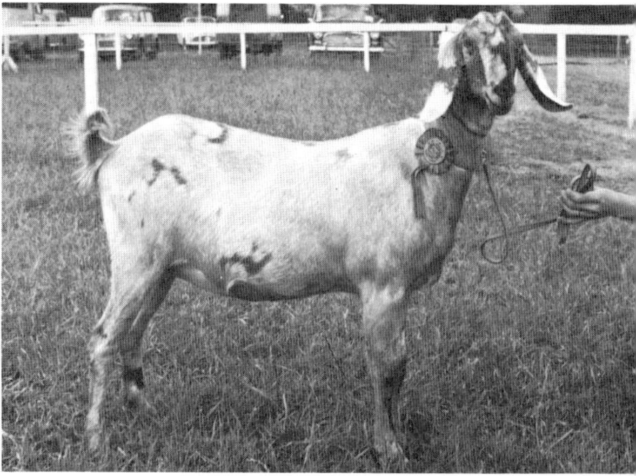

(*above*) Anglo-Nubian goatling

(*below*) Schwarzhal nannies and kids

some of these goats to England, where they became the property of Sir John Bagot. Since that time they have run as a semi-feral herd in Bagot's Park, Staffordshire, and have come to be known as the Bagot goat. In 1380 a goat's head was incorporated into the Bagot coat of arms, and the entrance lodge to Bagot's Park has a frieze of goat heads carved into the stonework.

The most distinctive feature of the Schwarzhal and Bagot goats is their colour. The head, neck and shoulders are jet black, while the remainder of the body is pure white. They have long hair, and both sexes are horned. The horns of the male are long and spreading, while those of the female are shorter and close together. A mature billy stands about 75 cm at the withers. The Bagot goats are allowed to breed freely, and most of the kids are born in February.

Swiss (*Saanen, Toggenberg, Alpine*)

Switzerland has been the source of most breeds of milk goat. Their milk has a higher fat and protein content than the milk of cattle, and it is useful for medical purposes as the fat globules are small, and the curds are softer and more easily digested. Goats will live on a wide variety of herbage, and a goat is often referred to as the poor man's cow. Because of its browsing habits it can be used to reclaim scrubland, but in some areas of the world excessive browsing by goats is creating desert conditions by killing the plant life. Milk goats spread from Switzerland to other parts of Europe, and from there to North America. The early settlers in Virginia took goats with them, but the first purebred stock was imported in 1893.

In Great Britain the British Goat Society is the parent organization which deals with general ad-

British Saanen nanny

ministration and publicity, but each breed has its own Society. Although many goats were imported originally from Switzerland, they have now become established as British breeds. The British Alpine, British Saanen and British Toggenberg each maintain a separate section within the British Goat Society's Herd Book.

The Toggenberg is the smallest of the most popular types. Mature females stand 65 cm at the withers and weigh 50 kg, while mature males weigh 65 kg. They are brown in colour with two white stripes down the face, white legs, and white marks on each side of the tail. The Alpine is a slimmer type so that while its weight is comparable to that of the Saanen, the mature females have a withers height of 77 cm. The colour is variable and ranges from white through brown and grey to black. The Saanen may be cream in colour but is usually white. The adult females have a withers height of 72 cm and weigh 55 kg, while the adult males weigh 75 kg. All types have a concave face and erect ears. They are short-haired, except for Alpine males which have a roach of long hair along the spine and a well-developed beard.

Although individual animals may produce in excess of 2500 kg of milk per lactation, the average yield is about 950 kg per lactation. The Toggenberg reaches a high peak of production in early lactation, but the yield of the Saanen is more persistent.

POULTRY

It is probable that all breeds of domestic poultry are descended from the Red Jungle Fowl, *Gallus gallus*, although some authorities claim that the mythical Gigantic Cock, *G. giganteus*, played a part in the evolution of such large breeds as the Cochin, Brahma and Malay Game. The range of the Red Jungle Fowl includes India and South-East Asia and, as its name implies, it is found in the jungles and is entirely absent from the alluvial plains and desert areas.

Domestic fowls first appeared in China around 1400 BC. They had spread into Greece and Phoenicia by 300 BC, and from there to western Europe and Britain by AD 100. Later they were taken by the Spanish Conquistadores and the Pilgrim Fathers to North America where several important and useful breeds have been evolved, such as the Wyandotte and the Plymouth Rock. The South American continent has its own breeds which are of early Asiatic origin. Aracaunas are of particular interest because they lay green and blue eggs.

Breeds of poultry can be created and developed, and then disappear relatively quickly. The three most popular breeds in the mid-20th century, the Leghorns, Rhode Island Red and Sussex, were unknown 100 years previously, and most of the popular breeds of that time are now almost extinct. This problem has been accentuated by the almost complete adoption of hybrid fowl by the poultry industry since the Second World War, and this has placed the majority of pure breeds in danger of extinction.

(*left*) Black Orpington cockerel

(*left*) White Leghorn hen

Fowls can be classified in various ways, but the most useful method is to group them according to their economic qualities. In general they can be divided into four groups or classes, namely: 1. Laying breeds 2. General-purpose breeds 3. Table breeds 4. Fancy breeds. Other methods of classification use colour of the plumage, or the division into 'light' and 'heavy' breeds, or the area of origin. Under the latter system there are basically two types, the Asiatic and the Mediterranean. The latter are small, active birds, with a large single comb, which are of little use for the table, but are excellent layers. They are rarely broody as eggs have been incubated artificially for centuries, since the time of the Egyptian egg ovens. Hens have been encouraged to continue laying rather than to hatch their eggs. The Asiatic fowls are large, heavy birds, suitable for the table and laying most of their eggs during the winter months.

The laying breeds produce the greatest number of eggs, but these are normally white-shelled and laid mainly in the spring and summer. Breeds in this category include the Ancona, Campine, Minorca, Leghorn, Redcap, Hamburgh, Scots Grey, Andalusian, Welsummer and Houdan.

General-purpose breeds combine both laying and table qualities, although their production standards do not match those of the specialist breeds for their particular quality. They are a useful fowl for keeping in the farmyard, as they produce tinted-shelled eggs with the main emphasis of production in the winter, and they are quite acceptable for the table afterwards. Breeds in this category include the Langshan, Orpington, Plymouth Rock, Wyandotte, Rhode Island Red, Light Sussex, North Holland Blue and Maran.

Light Sussex hen

The table breeds are large and deep-bodied, with a high proportion of the meat carried on the breast and wings. The chickens grow rapidly and fatten readily, but these breeds tend to be more delicate and some have yellow flesh which is not favoured by butchers and consumers. Breeds in this category include the Dorking, La Bresse, Faverolle, Indian Game and Crève Coeur.

Fancy breeds are by far the most numerous, as they include all game breeds, bantams, and freak breeds or deviations from standard types. Fancy breeds are selected primarily for their visual characteristics, such as comb, wattles, or plumage. The breeds included in this category are too numerous to list, but they range from the white Sultan from Turkey, whose face and small spiky comb are almost hidden by thick muffling, to the Transylvanian Naked Neck from Hungary, which has most of its neck entirely devoid of feathers.

Yokohama cock

(*above*) White Crested Black Polish cock

(*right*) Black Spanish cock

(*above*) Silver Campine cockerel

(*left*) White Malay cock

Turkeys, geese and ducks have followed a similar pattern of development, and are now largely produced by hybridization programmes. Turkeys are descended from the North American wild turkey, which favours a temperate forest habitat. Domestic geese trace their origin to the Grey-Lag Goose which occurs in many parts of Europe, Asia and northern Africa, but not in America. Domestic ducks are derived from the wild Mallard, although the Muscovy is an exception as it is descended from the wild Musk Duck from South America.

(*above right*)
American Bronze Turkey hen

(*below right*) White Turkey cockerel

(*left*) Brecon goose

(*right*) Embden gander

(*right*) Magpie drake

(*right*) Khaki drake

(*above*) Muscovy duck

(*left*) White Runner drake

BIBLIOGRAPHY

Anonymous, 1970. *British Sheep Breeds*. British Wool Marketing Board.

Briggs, H. M., 1958. *Modern Breeds of Livestock*. Macmillan.

Brynor Jones, C. (Editor), 1918. *Livestock of the Farm* (6 vols.). Gresham Publishing Co. Ltd.

Dobie, J. F., 1943. *The Longhorns*. Nicholson & Watson.

French, M. H. *et al.*, 1966. *European Breeds of Cattle* (2 vols.) F.A.O.

Garner, F. H., 1949. *The Cattle of Britain*. Longman.

Hicks, J. S. (Editor), 1921. *The Encyclopaedia of Poultry*. The Waverley Book Company.

Kelley, R. B., 1959. *Native and Adapted Cattle*. Angus & Robertson.

Layley, G. W. and Malden, W. J., 1935. *The Evolution of the British Pig*. J. Bale, Sons & Danielsson Ltd.

Low, D., 1842. *Domesticated Animals of the British Isles*. Longman.

Machin Goodall, D., 1965. *Horses of the World*. Country Life Ltd.

Mason, I. L., 1969. *A World Dictionary of Livestock Breeds, Types and Varieties*. Commonwealth Agricultural Bureaux.

M.A.F.F., 1938. *British Breeds of Livestock*. (Bulletin No. 86) H.M.S.O.

M.A.F.F., 1955. *Cattle of Britain*. (Bulletin No. 167) H.M.S.O.

M.A.F.F., 1960. *Sheep Breeding and Management*. (Bulletin No. 166) H.M.S.O.

M.A.F.F., 1960. *Poultry Breeding*. (Bulletin No. 146) H.M.S.O.

Quittet, E., 1963. *Les Races Bovines Françaises*. La Maison Rustique.

Quittet, E., 1965. *Les Races Ovines Françaises*. La Maison Rustique.

Quittet, E. and Portal M., 1963. *Les Races Porcines Françaises*. La Maison Rustique.

Quittet, E. and Richard P., 1963. *Les Races Chevalines Françaises*. La Maison Rustique.

Stuart, Lord David, 1970. *An Illustrated History of Belted Cattle*. Scottish Academic Press.

Thomas, J. F. H., 1948. *Sheep*. Faber.

Trow-Smith, R., 1959. *A History of British Livestock Husbandry* (2 vols.) Routledge & Kegan Paul.

Whitehead, G. K., 1972. *The Wild Goats of Great Britain and Ireland*. David & Charles.

Woods, Rex, 1976. *Rare Poultry of Ashe*. Spur Publishing Company.

Poultry Breed Societies. 1960.

British Poultry Standards. 1954. Poultry World, Ltd.

Journals

Focus on Beef (monthly). Richard D. Secord, 1108 Empire Building, Edmonton, Alberta, Canada.

Livestock International (bi-monthly). ACP Publishers Ltd., Wallingford-on-Thames, Oxfordshire, England.

The Ark (monthly). Countrywide Livestock Ltd., Eastrip House, Colerne, Chippenham, Wiltshire, England.

INDEX

Aberdeen Angus, 21, 93, 98, 109, 119, 126, 127
Afrikaner, 126, 136
Alpine (cattle), 16
Alpine (goat), 170
Ancona, 175
Andalusian (fowl), 175
Angeln, 119
Anglo-Nubian, 168
Arab, 22, 148, 149
Aracauna, 173
Ardennes (horse), 22
Argali, 17
Auroch, 16, 120
Awassi, 44
Axford, 166
Ayrshire, 94, 106, 130

'Babraham', 49, 77
Bagot, 170
Bakewell (Robert), 45, 46, 57, 59, 77, 120, 129
Bampton, 38
Beefmaster, 126
Beef Shorthorn, 119, 130, 132
Belgian, 22, 141, 153
Belted Galloway, 109
Berkshire, 23, 156, 162
Berkshire Knot, 49
Beulah Specklefaced, 38
Big Horn sheep, 17

Bison, 24
Blaarkop, 112
Blackheaded Persian, 44
Blanco Orejinegro, 99
Bleu du Maine, 25, 89
'Blue Cap', 86
Bluefaced Leicester, 20, 26, 30, 35, 81, 83
Blue-Grey, 110, 132
Border Leicester, 20, 21, 26, 28, 32, 34, 59, 60, 71, 83, 86
Brabançon, 141
Brahma, 173
Brahman, 96, 126
Brangus, 126
British Alpine, 171
British Lop, 160
British Saanen, 171
British Saddleback, 157
British Toggenburg, 171
British White, 16, 98, 139

Camargue cattle, 118
Cambridge, 34, 48, 62
Campine, 175
Cardy, 85
Celtic pony, 146, 147
Charbray, 100, 126
Charolais, 19, 21, 99, 119, 126, 129
Chester White, 158

Cheviot, 20, 28, 29, 30, 59, 70, 71, 90
Chianina, 21, 100
Chinese pigs, 23, 156, 164, 165
Chios, 44
Cleveland Bay, 142
Clun Forest, 20, 32, 56, 62, 66
Clydesdale, 22, 142, 144, 146, 152
Coates Herd Book, 129, 130
Cochin, 173
Colbred, 20, 34, 46, 83
Colburn (Oscar), 34, 46
Colling (Charles), 119, 129
Columbia, 64
Corriedale, 59, 64, 68
Cotswold, 36, 68, 91
Countrywide Livestock, 139
Crève Cœur, 176
Criollo (cattle), 120
Culley, 28
Cumberland, 159

Dairy Shorthorn, 129
Dalesbred, 82
Dales Pony, 22, 145
Danish Red, 124
Dartmoor (sheep), 38
Derbyshire Gritstone, 37, 90
Devon, 102, 104, 112, 135, 136
Devon Closewool, 20, 47
Devon Longwool, 47
Dexter, 104
Dishley Leicester, (see also Leicester), 28, 38, 40, 52, 59, 77, 82

Dorking, 176
Dorper, 44
Dorset Down, 20, 40, 54, 78
Dorset Horn, 18, 20, 21, 34, 41, 42, 64, 91
Drimmon, 119
Durham (cattle), 25, 129
Dutch Belted, 105

East Friesland, 18, 34, 40, 44, 59, 84
Eland, 24
Ellman (John), 46, 77
Equus przevalskii, 21
Essex, 157
Exmoor Horn, 20, 46, 64
Exmoor Pony, 22, 147

Faverolle, 176
Fell Pony, 146
Finnish Landrace, 19, 23, 44, 47
Fjall, 16, 98, 99
Fleckvieh, 133
Flemish (horse), 142, 144
Friesian, 16, 21, 96, 106, 122, 123, 127
Froment de Léon, 111
Fulani, 99

Galloway, 94, 98, 99, 106, 109, 132
Garne (W), 37
German Whiteheaded Mutton, 37
Gloucester, 110
Gloucester Old Spot, 159

188

Grey-Lag Goose, 180
Grey Steppe, 16, 102, 139
Guernsey, 21, 111

Hamburgh, 175
Hampshire (pig), 157
Hampshire Down, 20, 37, 40, 49, 78, 84
Herdwick, 50, 81
Hereford (cattle), 21, 112, 126
Hereford (sheep), 32, 57, 69
Highland, 16, 21, 114, 127, 138
Holderness, 129
Holstein, (see Friesian)
Houdan, 175
Humfrey (William), 49

Île de France, 26, 52, 59, 64
Indian Game, 176

Jacob, 44, 54
Jersey, 19, 21, 116, 127

Kent, (see Romney)
Kerry, 104, 117
Kerry Hill, 20, 56, 66
Kyloe, 114, 118

La Bresse, 176
Lacaune, 44
Lakenvelder, 105
Landrace (pig), 19, 23, 24, 160

Langshan, 175
Large Black, 161
Large White, 162
Leghorn, 173, 175
Leicester, (see also Dishley Leicester) 25, 31, 36, 37, 44, 56, 57, 60, 69, 73, 78, 84, 86, 91
Limestone, 37
Lincoln Longwool, 31, 57, 59, 64, 82, 84
Lincoln Red, 119, 129
Lincolnshire Curly Coat, 159
Linton, 70, 90
Llanllwni, 33
Llanwenog, 33
Lleyn, 59, 60, 84, 86
Longhorn, 120
Longmynd, 73
Lonk, 82, 90
Luing, 116

McCombie (William), 93
Maine-Anjou, 123, 129
Malay Game, 173
Mallard, 180
Manx Loghtan, 54
Maran, 175
Marchigiana, 102
Maremma, 102
Marsh sheep, 37, 44, 84
Merino, 19, 21, 31, 42, 53, 59, 62, 64, 74, 90
Meuse-Rhine-Ijssel, 21, 122, 130
Middle White, 23, 163
Minorca, 175
Montadale, 64
Morfe Common, 73

Moufflon, 17, 74
Mule (sheep), 20, 28, 69, 81
Muscovy, 180
Musk duck, 180

Navajo sheep, 55
Neapolitan pig, 23, 156
Norfolk Horn, 78
North Country Cheviot, 30, 68
Northern Dairy Shorthorn, 130
North Holland Blue, 175
North Ronaldsay, 18, 72
Nubian, 168

Oldenberg (horse), 144
Oldenberg (sheep), 37
Orpington, 175
Oryx, 24
Oxford Down, 20, 37, 80

Panama, 64
Penistone, 89
Percheron, 22, 149
Perendale, 32
Piedmont, 100
Pie Rouge de l'Est, 133
Pietrain, 19
Plymouth Rock, 173, 175
Poland China, 157, 165
Poll Dorset, 42, 43
Polwarth, 59
Portland, 18, 42, 64

Quarter Horse, 150

Radnor, 32, 66
Rambouillet, 52, 59, 63, 64
Red and White Canadian Holstein, 130
Red Angus, 94, 120
Red Barbadan, 23, 166
Redcap, 175
Red deer, 24
Red Jungle Fowl, 173
Red Poll, 98, 124
Rhiw, 85
Rhode Island Red, 173, 175
Romagnola, 100
Romanov, 30, 48
Romans, 20, 62
Romedale, 64
Romney, 32, 66, 78, 84
Roscommon, 60
Rouge de l'Ouest, 26
Rough Fell, 82
Ryeland, 20, 31, 32, 42, 69, 78

Saanen, 170
St Kilda, 54
St Rona's Hill, 54
Santa Gertrudis, 21, 124
Schwarzhal, 168
Scots Grey, 175
Scottish Blackface, 20, 29, 30, 31, 70, 90
Sennybridge Cheviot, 30
Sheeted Somerset, 106
Shetland (cattle), 127
Shetland (sheep), 18, 72
Shire, 142, 145, 152
Shorthorn, 21, 94, 99, 111, 119, 120, 123, 126, 127, 128

190

Shropshire, 20, 32, 33, 73, 78, 80
Simmental, 21, 99, 130, 132
Small White, 163
Soay, 17, 18, 23, 24, 72, 74
South Devon (cattle), 134
South Devon (sheep), 38
Southdown, 20, 40, 42, 49, 50, 68, 69, 73, 74, 77, 79, 91
Suffolk (horse), 153
Suffolk (sheep), 20, 41, 50, 74, 78, 84
Suffolk Dun, 98, 124
Sussex (cattle), 112, 135
Sussex (fowl), 173, 175
Sultan, 177
Swaledale, 20, 52, 71, 81, 90
Swiss goats, 170

Tamworth, 23, 24, 166
Targhee, 64
Tarpan, 22
Teeswater (cattle), 119, 129, 130
Teeswater (sheep), 27, 28, 40, 59, 81, 82, 86, 89
Texas Longhorn, 24, 99, 126
Texel, 26, 59, 84
Thoroughbred, 22, 144, 150
Toggenberg, 170
Tompkins (Benjamin), 112

Transylvanian Naked Neck, 177

Urial, 17

Vikings, 16, 50, 98, 109, 116

Watson (Hugh), 93
Webb (Jonas), 49
Welsh (pig), 160
Welsh Black, 118, 136
Welsh Mountain, 20, 29, 59, 60, 64, 85, 92
Welsummer, 175
Wensleydale, 25, 27, 59, 83, 86
Wessex, 157
Whitebred Shorthorn, 110, 132
Whitefaced Woodland, 64, 89
White Park, 98, 138, 139
Wild boar, 17, 23
Wiltshire Horn, 49, 59, 90
Wyandotte, 173, 175

Yorkshire (cattle), 129
Yorkshire (pig), 158, 163
Yorkshire Coach Horse, 143
Young (Arthur), 78